AF661481

BIOTECHNOLOGY AND THE FUTURE OF FOOD

BIOTECHNOLOGY AND THE FUTURE OF FOOD

Kamal Duggal

CYBER TECH PUBLICATIONS
4264/3, Ansari Road, Daryaganj, New Delhi-110002 (India)
Ph.: 011-23244078. Fax: 011-23280028
E-Mail: cyberpublicationsdelhi@yahoo.com

BIOTECHNOLOGY AND THE FUTURE OF FOOD

Kamal Duggal

© Reserved

First Edition 2009

All rights reserved. No part of this publication may be reproduced, stored in a retrieval system or transmitted, in any form or by any means, electronic, mechanical, photo copying, recording or otherwise, without the prior permission of the publisher.

Published by :

G.S. Rawat for **Cyber Tech Publications**
4264/3, Ansari Road, Daryaganj, New Delhi-110002 (India)
Ph.: 011-23244078. Fax: 011-23280028
E-Mail: cyberpublicationsdelhi@yahoo.com

Printed at : **Amit Enterprises**, Maujpur, Delhi, (9810878184)

Preface

The need to identify biodiversity research priorities was identified as a priority action in the National Strategy for the Conservation of Biodiversity.

A high level National Biodiversity Research Priorities Strategic Planning Workshop was held on 10 April 2000. This workshop brought together experts in a broad range of biodiversity conservation fields including ecology, taxonomy, land and marine area management, research and development, informatics and economics. The States, regional and catchment management authorities, local government, industry and the community were also represented.

Some areas of research are required by legislation such as the Environment Protection and Biodiversity Conservation Act 1999. This Act focuses Commonwealth interests on matters of national environmental significance, puts in place a streamlined environmental assessment and approvals process, and establishes an integrated regime for biodiversity conservation and the management of protected areas. The Act recognises the responsibility of the States for delivering on-ground natural resource management. Two of the objectives of the Act are 'to promote the conservation of biodiversity' and 'to promote ecologically sustainable development through the conservation and ecologically sustainable use of natural resources'.

— *Author*

CONTENTS

1

INTRODUCTION

It is safe to say that one thing you'll do today is eat some food--food is pretty important to all animals. If you don't eat, it can cause all sorts of problems: hunger, weakness, starvation... Food is essential to life.

But what is food? What's in food that makes it so important? What happens to the food once you eat it? What is food made of? How does it fuel our bodies? What do words like "carbohydrates" and "fats" really mean (especially on those "Nutrition Facts" labels you find on almost everything these days)? What would happen if you ate nothing but marshmallows for a week? What is a calorie? Why can't we eat grass like a cow does, or wood like a termite?

If you have ever wondered about food and how your body uses it, then read on. In this article, we'll give you all of the information you need to understand what a hamburger or a banana does to keep your body running every day.

THE BASICS OF FOOD

Think about some of the things you have eaten today--maybe cereal, bread, milk, juice, ham, cheese, an apple, potatoes... All of these foods (and pretty much any other food that you can think of) contain seven basic components:

- Carbohydrates (simple and complex)
- Proteins
- Fats

- Vitamins
- Minerals
- Fiber
- Water

Your body's goal is to digest food and use it to keep your body alive. In the following sections, we will look at each of these basic components to understand what they really do and why they are so important to your body.

(Note that there might be a few non-food things mixed in with what you eat, especially if you are eating lots of processed foods. Things like artificial colors and chemical preservatives are the most common. Those are additives, not part of the natural foods.)

CARBOHYDRATES

You have probably heard of "carbohydrates" and "complex carbohydrates." Carbohydrates provide your body with its basic fuel. Your body thinks about carbohydrates like a car engine thinks about gasoline.

The simplest carbohydrate is glucose. Glucose, also called "blood sugar" and "dextrose," flows in the bloodstream so that it is available to every cell in your body. Your cells absorb glucose and convert it into energy to drive the cell. Specifically, a set of chemical reactions on glucose creates ATP (adenosine triphosphate), and a phosphate bond in ATP powers most of the machinery in any human cell. If you drink a solution of water and glucose, the glucose passes directly from your digestive system into the bloodstream.

The word "carbohydrate" comes from the fact that glucose is made up of carbon and water. The chemical formula for glucose is:

You can see that glucose is made of six carbon atoms (carbo...) and the elements of six water molecules (...hydrate). Glucose is a simple sugar, meaning that to our tongues it tastes sweet. There are other simple sugars that you have probably heard of. Fructose is the main sugar in fruits. Fructose has the same chemical formula as glucose ($C6H_{12}O_6$), but the atoms are arranged slightly differently. The liver converts fructose to glucose. Sucrose, also known as "white sugar" or "table sugar," is made of one glucose and one fructose molecule bonded together. Lactose (the sugar found in milk) is made of one glucose and one galactose molecule bonded together. Galactose, like fructose, has the same chemical components as glucose but the atoms are arranged differently. The liver also converts galactose to glucose. Maltose, the sugar found in malt, is made from two glucose atoms bonded together.

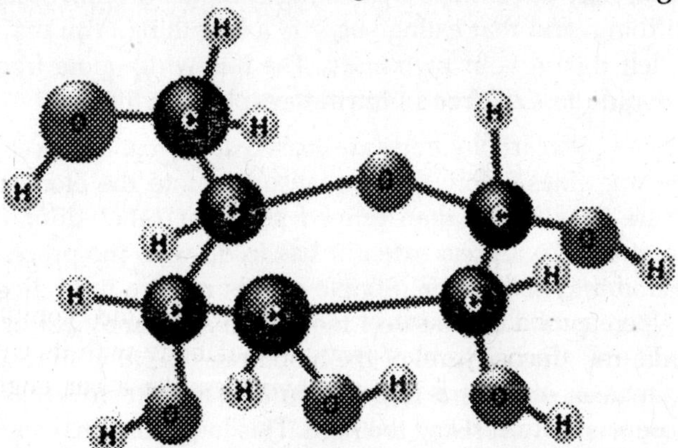

Glucose

Glucose, fructose and galactose are monosaccharides and are the only carbohydrates that can be absorbed into the bloodstream through the intestinal lining. Lactose, sucrose and maltose are disaccharides (they contain two monosaccharides) and are easily converted to their monosaccharide bases by enzymes in the digestive tract. Monosaccharides and disaccharides are called simple carbohydrates. They are also sugars--they all taste sweet. They all digest quickly and enter the bloodstream quickly. When you look at a "Nutrition Facts" label on a food package and see "Sugars" under the "Carbohydrates" section of the label, these simple sugars are what the label is talking about.

There are also complex carbohydrates, commonly known as "starches." A complex carbohydrate is made up of chains of glucose molecules. Starches are the way plants store energy--plants produce glucose and chain the glucose molecules together to form starch. Most grains (wheat, corn, oats, rice) and things like potatoes and plantains are high in starch. Your digestive system breaks a complex carbohydrate (starch) back down into its component glucose molecules so that the glucose can enter your bloodstream. It takes a lot longer to break down a starch, however. If you drink a can of soda full of sugar, glucose will enter the bloodstream at a rate of something like 30 calories per minute. A complex carbohydrate is digested more slowly, so glucose enters the bloodstream at a rate of only 2 calories per minute (reference).

You may have heard that eating complex carbohydrates is a good thing, and that eating sugar is a bad thing. You may even have felt this in your own body. The following quote from The Yale Guide to Children's Nutrition explains why:

If complex carbohydrates are broken down to monosaccharides in the intestines before they are absorbed into the bloodstream, why are they better than refined sugar or other di-or mono-saccharides? To a great extent it has to do with the processes of digestion and absorption. Simple sugars require little digestion, and when a child eats a sweet food, such as a candy bar or a can of soda, the glucose level of the blood rises rapidly. In response, the pancreas secretes a large amount of insulin to keep blood glucose levels from rising too high. This large insulin response in turn tends to make the blood sugar fall to levels that are too low 3 to 5 hours after the candy bar or can of soda has been consumed. This tendency of blood glucose levels to fall may then lead to an adrenaline surge, which in turn can cause nervousness and irritability... The same roller-coaster ride of glucose and hormone levels is not experienced after eating complex carbohydrates or after eating a balanced meal because the digestion and absorption processes are much slower.

If you think about it, this is incredibly interesting because it shows that the foods you eat and the way you eat them can affect your mood and your temperament. Foods do that by affecting the levels of different hormones in your bloodstream over time. Another interesting thing about this quote is its mention of insulin. It turns out that insulin is incredibly important to the way the body uses the glucose that foods provide. The functions of insulin are:

- To enable glucose to be transported across cell membranes
- To convert glucose into glycogen for storage in the liver and muscles

- To help excess glucose be converted into fat
- To prevent protein breakdown for energy

According to the Encyclopedia Britannica:

Insulin is a simple protein in which two polypeptide chains of amino acids are joined by disulfide linkages. Insulin helps transfer glucose into cells so that they can oxidize the glucose to produce energy for the body. In adipose (fat) tissue, insulin facilitates the storage of glucose and its conversion to fatty acids. Insulin also slows the breakdown of fatty acids. In muscle it promotes the uptake of amino acids for making proteins. In the liver it helps convert glucose into glycogen (the storage carbohydrate of animals) and it decreases gluconeogenesis (the formation of glucose from noncarbohydrate sources). The action of insulin is opposed by glucagon, another pancreatic hormone, and by epinephrine.

What you can begin to see from this description is that there are actually lots of different things happening in your body around glucose. Because glucose is the essential energy source for your body, your body has many different mechanisms to ensure that the right level of glucose is flowing in the bloodstream. For example, your body stores glucose in your liver (as glycogen) and can also convert protein to glucose if necessary. Carbohydrates provide the energy that cells need to survive.

PROTEINS

A protein is any chain of amino acids. An amino acid is a small molecule that acts as the building block of any cell. Carbohydrates provide cells with energy, while amino acids provide cells with the building material they need to grow and maintain their structure. Your body is about 20-percent protein by weight.

It is about 60-percent water. Most of the rest of your body is composed of minerals (for example, calcium in your bones). Amino acids are called "amino acids" because they all contain an amino group (NH_2) and a carboxyl group (COOH), which is acidic. Below you can see the chemical structure of two of the amino acids.

You can see that the top part of each is identical to the other. That is true of all amino acids--the little chain at the bottom (the H or the CH_3 in these two amino acids) is the only thing varying from one amino acid to the next. In some amino acids, the variable part can be quite large. The human body is constructed of 20 different amino acids (there are perhaps 100 different amino acids available in nature).

Two of the Amino Acids

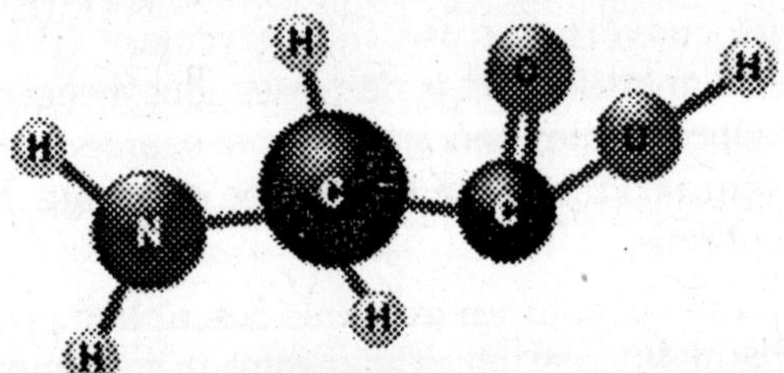

Glycine

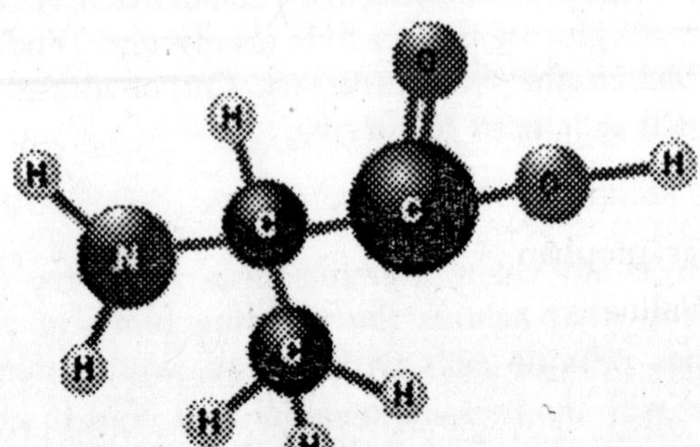

Alanine

As far as your body is concerned, there are two different types of amino acids: essential and non-essential. Non-essential amino acids are amino acids that your body can create out of other chemicals found in your body. Essential amino acids cannot be created, and therefore the only way to get them is through food. Here are the different amino acids:

Non-essential

- Alanine (synthesized from pyruvic acid)
- Arginine (synthesized from glutamic acid)
- Asparagine (synthesized from aspartic acid)
- Aspartic Acid (synthesized from oxaloacetic acid)
- Cysteine
- Glutamic Acid (synthesized from oxoglutaric acid)
- Glutamine (synthesized from glutamic acid)
- Glycine (synthesized from serine and threonine)
- Proline (synthesized from glutamic acid)
- Serine (synthesized from glucose)
- Tryosine (synthesized from phenylalanine)

Essential

- Histidine
- Isoleucine
- Leucine
- Lysine
- Methionine
- Phenylalanine
- Threonine
- Tryptophan
- Valine

Protein in our diets comes from both animal and vegetable sources. Most animal sources (meat, milk, eggs) provide what's called "complete protein," meaning that they contain all of the essential amino acids. Vegetable sources usually are low on or missing certain essential amino acids. For example, rice is low in isoleucine and lysine. However, different vegetable sources are deficient in different amino acids, and by combining different foods you can get all of the essential amino acids throughout the course of the day. Some vegetable sources contain quite a bit of protein--things like nuts, beans, soybeans, etc. are all high in protein. By combining them you can get complete coverage of all essential amino acids. The digestive system breaks all proteins down into their amino acids so that they can enter the bloodstream. Cells then use the amino acids as building blocks.

From this discussion you can see that your body cannot survive strictly on carbohydrates. You must have protein. According to this article, the RDA (Recommended Daily Allowance) for protein is 0.36 grams of protein per pound of body weight. So a 150-pound person needs 54 grams of protein per day. The photo above is the Nutritional Facts label from a can of tuna. You can see that a can of tuna contains about 32 grams of protein (this can has 13 grams per serving and there are 2.5 servings in the can). A glass of milk contains about 8 grams of protein. A slice of bread might contain 2 or 3 grams of protein. You can see that it is not that hard to meet the RDA for protein with a normal diet.

FATS

We all know about the common fats that different foods contain. Meat contains animal fat. Most breads and pastries contain vegetable oils, shortening or lard. Deep fried foods are cooked in heated oils. Fats are greasy and slick.

You commonly hear about two kinds of fats: saturated and unsaturated. Saturated fats are normally solid at room temperature, while unsaturated fats are liquid at room temperature. Vegetable oils are the best examples of unsaturated fats, while lard and shortening (along with the animal fat you see in raw meat) are saturated fats. However, most fats contain a mixture. For example, above you see the label from a bottle of olive oil. It contains both saturated and unsaturated fats, but the saturated fats are dissolved in the unsaturated fats. To separate them, you can put olive oil in the refrigerator. The saturated fats will solidify and the unsaturated fats will remain liquid. You can see that the olive oil bottler even chose to further distinguish the unsaturated fats between polyunsaturated and monounsaturated. Unsaturated fats are currently thought to be more healthy than saturated fats, and monounsaturated fats (as found in olive oil and peanut oil) are thought to be healthier than polyunsaturated fats.

Fats that you eat enter the digestive system and meet with an enzyme called lipase. Lipase breaks the fat into its parts: glycerol and fatty acids. These components are then reassembled into triglycerides for transport in the bloodstream. Muscle cells and fat (adipose) cells absorb the triglycerides either to store them or to burn them as fuel. You need to eat fat for several reasons:

- As we will see in the next section, certain vitamins are fat soluble. The only way to get these vitamins is to eat fat.
- In the same way that there are essential amino acids, there are essential fatty acids (for example, linoleic acid is used to build cell membranes). You must obtain these fatty acids from food you eat because your body has no way to make them.
- Fat turns out to be a good source of energy. Fat contains twice as many calories per gram as do carbohydrates or proteins. Your body can burn fat as fuel when necessary.

For more information about fat in the diet, check out the links page at the end of this article.

VITAMINS

The Merriam-Webster Collegiate Dictionary defines "vitamin" as: Any of various organic substances that are essential in minute quantities to the nutrition of most animals and some plants, act esp. as coenzymes and precursors of coenzymes in the regulation of metabolic processes but do not provide energy or serve as building units, and are present in natural foodstuffs or sometimes produced within the body.

Vitamins are smallish molecules (Vitamin B12 is the largest, with a molecular weight of 1,355) that your body needs to keep itself running properly. In How Sunburns and Sun Tans Work, we learn that the body can produce its own Vitamin D, but generally vitamins must be provided in food. The human body needs 13 different vitamins:

- Vitamin A (fat soluble, retinol) comes from beta-carotene in plants; when you eat beta-carotene, an enzyme in the stomach turns it into Vitamin A.

- Vitamin B (water soluble, several specific vitamins in the complex)
- Vitamin B1: Thiamine
- Vitamin B2: Riboflavin
- Vitamin B3: Niacin
- Vitamin B6: Pyridoxine
- Vitamin B12: Cyanocobalamin
- Folic Acid
- Vitamin C (water soluble, ascorbic acid)
- Vitamin D (fat soluble, calciferol)
- Vitamin E (fat soluble, tocopherol)
- Vitamin K (fat soluble, menaquinone)
- Pantothenic acid (water soluble)
- Biotin (water soluble)

In most cases, the lack of a vitamin causes severe problems. The following list shows diseases associated with the lack of different vitamins:

- Lack of Vitamin A: Night blindness, xerophthalmia
- Lack of Vitamin B1: Beriberi
- Lack of Vitamin B2: Problems with lips, tongue, skin,
- Lack of Vitamin B3: Pellagra
- Lack of Vitamin B12: Pernicious anemia
- Lack of Vitamin C: Scurvy
- Lack of Vitamin D: Rickets
- Lack of Vitamin E: Malabsorption of fats, anemia
- Lack of Vitamin K: Poor blood clotting, internal bleeding

A diet of fresh, natural food usually provides all of the vitamins that you need. Processing tends to destroy vitamins, so many processed foods are "fortified" with man-made vitamins.

MINERALS

Minerals are elements that our bodies must have in order to create specific molecules needed in the body. Here are some of the more common minerals our bodies need:

- Calcium-used by teeth, bones
- Chlorine
- Chromium
- Copper
- Fluorine-strengthens teeth
- Iodine-combines with tryosine to create the hormone thyroxine
- Iron-transports oxygen in red blood cells
- Magnesium
- Manganese
- Molybdenum
- Phosphorus
- Potassium-important ion in nerve cells
- Selenium
- Sodium
- Zinc

We do need other minerals, but they are supplied in the molecule that uses them. For example, sulfur comes in via the amino acid methionine, and cobalt comes in as part of vitamin B12.

Food provides these minerals. If they are lacking in the diet, then various problems and diseases arise.

WATER

As mentioned above, your body is about 60-percent water. A person at rest loses about 40 ounces of water per day.

Water leaves your body in the urine, in your breath when you exhale, by evaporation through your skin, etc. Obviously, if you are working and sweating hard then you can lose much more water. Because we are losing water all the time, we must replace it.

We need to take in at least 40 ounces a day in the form of moist foods and liquids. In hot weather and when exercising, your body may need twice that amount. Many foods contain a surprising amount of water, especially fruits. Pure water and drinks provide the rest.

FIBERS

Fiber is the broad name given to the things we eat that our bodies cannot digest. The three fibers we eat on a regular basis are:

- Cellulose
- Hemicellulose
- Pectin

Hemicellulose is found in the hulls of different grains like wheat. Bran is hemicellulose. Cellulose is the structural component of plants.

It gives a vegetable its familiar shape. Pectin is found most often in fruits, and is soluble in water but non-digestible. Pectin is normally called "water-soluble fiber" and forms a gel. When we eat fiber, it simply passes straight through, untouched by the digestive system.

Cellulose is a complex carbohydrate. It is a chain of glucose molecules. Some animals and insects can digest cellulose. Both cows and termites have no problem with it because they have bacteria in their digestive systems secreting enzymes that break down cellulose into glucose.

Human beings have neither the enzymes nor these beneficial bacteria, so cellulose is fiber for us.

I'm Starving...

A normal person who is eating three meals a day and snacking between meals gets almost all of his or her energy from the glucose that carbohydrates provide. What happens if you stop eating, however?

For example, what if you are lost in the woods, or you are purposefully fasting? What does your body do for energy? Your body goes through several phases in its attempt to keep you alive in the absence of food.

The first line of defense against starvation is the liver. The liver stores glucose by converting it to glycogen. It holds perhaps a 12-hour supply of glucose in its glycogen.

Once you finish digesting all of the carbohydrates that you last ate, the liver starts converting its stored glycogen back into glucose and releases it to maintain glucose in the blood. Lipolysis also starts breaking down fat in the fat cells and releasing fatty acids into the bloodstream.

Tissues that do not need to use glucose for energy (for example, muscle cells) start burning the fatty acids. This reduces the glucose demand so that nerve cells get the glucose. Once the liver runs out of glycogen, the liver converts to a process called gluconeogenesis. Gluconeogenesis turns amino acids into glucose.

The liver then begins producing ketone bodies from fatty acids being made available in the blood by lipolysis. Brain and nerve cells convert over from being pure consumers of glucose to partial consumers of ketone bodies for energy.

Some of these alternative metabolic processes are actually used on a regular basis. For example, Eskimos eating a traditional Eskimo diet have virtually no carbohydrates on the menu.

You may have also read about several recent weight-loss programs that try to take advantage of ketone metabolism to "burn fat" (this article offers a thorough description of the "ketogenic diet" as used in medicine, and this article talks about the "fad diets" that utilize the ketone effect).

When you hear about these diets you will now have a better idea of what they're about!

THERAPEUTIC NUTRITION

Nutritional supplementation is the addition of selected nutrients in amounts higher than in the current diet.

The body has mechanisms to control the absorption, storage and excretion of most nutrients, this results in a wide margin of safety. Even so, the amount of a particular nutrient supplemented should be based on age, weight, reproductive status, metabolic and disease situation of the individual and not a one size fits all recommendation. Proper nutritional supplementation can greatly enhance both therapeutic and preventive health regimens.

Practitioners should participate heavily in the design of supplementation programs for all of their patients to whom nutrients are supplemented, however, too many products are being introduced to maintain current knowledge of them all. Products need to be evaluated on an ingredient by ingredient basis.

The following paper will give guidelines on what questions to ask, how to gather answers, how to compare product to product, and will list some nutrients commonly supplemented with brief descriptions and grouped according to common uses. Herbal ingredients will not be discussed. Herbs are not nutrition, they contain chemicals that exert pharmacological activities.

DO THEY WORK?

A very common question that equine practitioners are asked about supplements is, does this product work? To answer that question, work must be defined. Relief of pain or other symptoms may be the desired end-point. If so, short term supplementation of many ingredients including herbs and hormone pre-cursors will work. Most of these types of ingredients should not be supplemented for long periods. As opposed to chemicals that enter the body and change how the metabolic processes work, supplementation of nutrients provide extra raw materials that are incorporated into the normal functioning of a system. Except in the case of a frank deficiency, improvements seen through nutritional supplementation are much slower than the changes caused by drug therapies. Many times the enzyme systems must "ramp up" before they can utilize the newly provided nutrients. Depending on the turnover rate of the tissues in question, structural changes through nutritional supplementation can take weeks, not days, to be seen. If the desired end-point is correction of an underlying problem, not just treatment of symptoms, then there are fewer products that actually work.

HOW TO GATHER INFORMATION

The first step in categorizing products or ingredients between the Good, the Bad and the Useless is to gather reliable information. As with any other product, marketing literature is a good source of potential uses, but it can't be taken at face value. Independent research reported in peer reviewed journals is the most trustworthy source of information, followed by case series and case studies. However, this type of independent verification is much more rare in the supplement industry than others because of the difficulty in obtaining and defending patents. With the decreased ability to protect proprietary products comes decreased profits and decreased money for research.

Much more information is available on individual nutrients than complete products. When reviewing the current literature on nutrients, attention must be paid to the specific forms, dose, routes of administration, and measures of change used in the studies. There can be much more variation in expected activity based on these specifics than on which species was used in the study. If the ingredients in question are basic nutrients, extrapolation of results from one species to another can be reasonable.

WHAT IS BEHIND THE LABEL?

Evaluation of commercial products can be very difficult. The nutritional supplement industry is very different from the pharmaceutical industry. There are several small companies instead of a few large ones. The availability of technical support by professionals with appropriate credentials is very scarce. Selection of a company can be as important as the selection of what nutrients to supplement. When you read a company's literature was it written by the marketing department or were there people that know the science involved? When you call a company can you readily speak with a professional in technical support? Beware of "one size fits all" doses. The amount of nutrients needed by an individual is based on weight, age, and severity of the problem or deficiency. A company should be able to quickly help you use their product in a way customized to your patient. While ingredient quality is of utmost importance in a supplement, it is very difficult to verify. Unfortunately, a practitioner's clinical experiences and personal interactions with a company will yield the best indications of product quality.

GENERAL SUPPLEMENTS

Antioxidants

Vitamins, minerals, and other molecules, which act in the body to scavenge oxygen radicals, are antioxidants. Most patients that need nutritional support will benefit from a high level of dietary antioxidants.

Antioxidants should be supplemented as a group instead of high levels of individual ones, due to the fact that any antioxidant can act as a pro-oxidant if present in high enough amounts.

Minerals

Minerals act as cofactors for enzymes for almost every reaction in the body. Everything from immune system function, to bone density, to protein, fat and CHO metabolism is affected by mineral deficiencies. Mineral supplementation should be critically evaluated.

The authors suggest amino acid chelated minerals, due to the increased bioavailability of these formulations. Other organic complexes (citrates, gluconates, and lactates) have a higher biological value than the inorganic complexes (oxides, carbonates) with sulfates being in the middle. The inorganic forms of many minerals are toxic, while the organic forms are very beneficial (ex: chromium). Unless minerals are supplied in the correct balance and form, minimal benefit will be obtained by supplementation.

Vitamins

Vitamins are categorized into two distinct classes: water soluble and fat soluble. Water soluble vitamins are not stored in the body; they are excreted if not utilized soon after ingestion. Fat soluble vitamins (Vitamins A, D, E, and K) are stored and have more of a potential for toxicity. Horses on premium commercial feeds rarely develop a clinical deficiency of an individual mineral or vitamin because the feeds are appropriately fortified. General supplements should not be used to try to make a good ration out of cheap feedstuffs. They can be used to support horses that are under more stressful conditions such as performance, reproductively active, diseased, or geriatric.

Hoof

Hoof health is often a reflection of the nutritional status of the individual. Keratin based structures such as the hoof, mane, tail and hair may reflect nutritional deficiencies long before other body systems begin to exert obvious signs. Vitamins, minerals, and specific amino acids maybe spared incorporation into these structures due to the need in more biologically "important" structures. However some individuals may need supplementation for the hoof and other keratin based structures beyond what can be supplied in even premium horse feeds. In these cases attention should be given to specific nutrients that are necessary for proper keratin formation. Guidelines provided in this article for evaluation of any supplement should be applied when choosing a "hoof supplement". In addition the following is a list and short description of specific nutrients that provide the body with the basic building blocks of keratin.

Biotin: Biotin is a B vitamin that is required by all animals as a biological catalyst in numerous metabolic reactions. Biotin is involved in the keratinization process early on in the horn cell's development. Due to its early incorporation, the benefit of biotin supplementation is often not seen for months or until new hoof growth is evident. Animals with biotin deficiency present with dry and cracked skin, poor hair coats, split and cracked hooves and foot lesions. Though biotin may be the most well known "hoof nutrient" it is not the only one. Some horses supplemented with biotin alone exhibit no improvement in overall hoof health. This often suggest a deficiency in other nutrients that are integral in healthy hoof development.

Amino Acids: Lysine and Methionine are two amino acids essential for quality hoof growth. Keratin is a protein. Proteins are huge complex molecules that involve numerous amino acids for the complete formation. Proteins that lack enough of the correct types of amino acids often are not functionally sound, resulting in poor quality hooves. Methionine and Lysine are highly utilized in the horse and should be supplied in adequate amounts in the diet.

Joint

It was found over 50 years ago, that over time, eating ground up connective tissues could help relieve problems associated with arthritis. People began to consume gelatin and cartilage (shark, chicken, bovine, or Perna muscle) and these helped some. Studies found that chondroitin sulfates were contributing most of the beneficial effects. About 30 years ago the research began focusing in on the much smaller Glucosamine Sulphate.

Whole cartilage and its large components (chondroitin sulfates and collagen) can not be absorbed well by the body. While large molecules such as these can be injected and work very well, when taken orally more than half usually ends up in the manure. Chondroitin sulfates have a large range in size. Some are 50 times larger than others. Only 8-10% of the smallest ones have been shown to get from the gut into the blood intact. What is fed must be digested by the gut and the pieces that do get into the blood must be broken down further to get into the joints.

There are several types of glucosamine used in supplements.

Glucosamine Sulphate is the basic building block of connective tissues and fluids. Studies have shown in humans, rats, and dogs, that when Glucosamine Sulphate is given orally, within 30 minutes 87-97% is actively taken from the gut into the blood. Further, within 4 hours the chondrocytes (joint cells) have actively taken it from the blood. When linked together with sulfur, chains of glucosamines become several different types of connective tissues and joint fluid. [A lack of sulfur will cause the production of connective tissues to stop.] Also an enzyme can slightly change the shape of the glucosamine into galactosamine, the basic building block of the chondroitin sulfates. In one clinical study, where Glucosamine Sulphate was feed to horses, 77% of Navicular disease and 100% of Spavin cases returned to normal function. N-acetyl-glucosamine has been proven to not have active uptake from the gut in vitro.

There are no studies that show if Glucosamine HCl is actively absorbed, or how much of it actually gets into the blood or the joints. Whatever portion does make it into the joints must be undergo changes before it can be used in connective tissues. The HCl must be removed and a sulphate added. Furthermore, Glucosamine HCl is not stable in liquids. Published experiments have shown that over half of the Glucosamine HCl added to a liquid solution will breakdown within 27 hours. Within 4 days, these new and different "breakdown products" form completely different molecules that were 20 times larger than what was originally in the solution.

Other non-structural ingredients are also commonly used.

Glutamic acid (glutamate) is an excitatory amino acid (EAA). When this amino acid is added to a brain experimentally it causes seizures. Higher than normal concentrations are present in the brain of human seizure patients. Research is being done to see whether blocking its activity will help control seizures. It is also being considered as an antidepressant drug. This amino acid is also naturally released by joint nerves to signal pain. It is used as an experimental marker to determine the amount of pain induced in experimental models.

The more glutamate that is released, the more pain that is felt in a joint. Caution is given here because a review of this product in a popular publication discusses its dramatic and swift change in horses' attitudes. Research is currently being done to determine its role in seizures and the signaling of pain. What will prolonged use do?

At what levels does this ingredient become dangerous? How addicting is it? These are only some of the questions not clearly answered by science. Glutamic acid is added to some joint supplements because the amine group in glucosamine can be donated by glutamic acid. However, providing extra of a particular AA does not mean that more connective tissue will be made.

Bromelain is an enzyme found in pineapples. It has been shown to reduce the amount of swelling present in some experimental animals. However, Bromelain can enter the body intact, because it digests the lining of the GI tract. In human cancer patients, it has been shown to stimulate the body's own immune system to kill somatic cells. Since it is a protein, it can cause allergic reactions. People who handle bromelain have become allergic to it.

Many of the "flex products" on the market contain herbs, which do not contribute nutrients to the joint tissues. While an in depth discussion of herbal ingredients is beyond the scope of this paper, the following are mentioned due to their common addition to many supplements and their serious impact on or conflict with other drugs and their potential to cause positive results on competitive drug screenings.

Yucca contains steroid saponins. Steroids have been shown to slow the production of glycosaminoglycans (GAG's). Boswellia, white willow's bark, and snake root are all herbs containing chemicals that act as NSAID's. Devil's claw contains several chemicals, which are reported to decrease pain. They do not decrease inflammation and are similar in structure to steroids. This herb should not be used with any female (human or animal), as it has been reported that Devil's claw can cause abortions, by stimulating uterine contractions.

ELECTROLYTES

Electrolytes are not stored in the body therefore daily requirements must be met through the diet. The major electrolytes are sodium, chloride and potassium. The voluntary intake of electrolytes is balanced by the output of electrolytes in urine, feces, and sweat in normal healthy horses. Dietary loads of electrolytes increase the urine losses of these electrolytes.

However, in certain situations such as strenuous exercise, diarrhea, and other gastrointestinal abnormalities; the electrolyte losses may exceed the intake. In these situations, supplementation with salt (NaCl) and Lite salt (KCl) may be necessary to offset these losses. Horses with profuse diarrhea often are affected by electrolyte imbalances, however, these needs are normally corrected by intravenous fluid therapy. Sodium bicarbonate supplementation is also necessary in some disease states and may be beneficial in exercising horses to buffer the production of lactic acid.

Horses that exercise for long periods of time and become depleted of electrolytes often benefit from oral supplementation. Not only does supplementation correct the electrolyte levels, it also stimulates the horse to drink more water. When choosing an electrolyte supplement, use a trustworthy manufacturer and make sure you are supplementing the electrolyte minerals, not just sugars.

FATS AND OILS

Fats and oils make up a family of chemical compounds called lipids. Lipids that are solid at room temperature are called fats. Lipids that are liquid at room temperature are called oils. Lipids are capable of providing large amounts of energy (9 kcal/gram), twice as much as protein and carbohydrates on a per gram basis (4 kcal/gram).

For this reason it is often advantageous to supplement lipids in the form of oils when an individual is in need of extra energy. Less heat is produced as a by-product of fat digestion. This is beneficial in climates with high heat and humidity. In equine rations, animal sources of fats, such as tallow, are between 88 and 92 percent digestible; whereas plant sources, such as corn or soybean oils, are around 94 percent digestible. Corn oil is the most palatable lipid supplement for horses.

Fat should be added to the diet gradually (¼ cup per meal initially and increase by ¼ cup every three days). Horses may require up to 6 weeks to adjust their metabolism to the increased fat utilization. Higher fat diets (20-25%) have been shown to enhance fatty acid oxidation in horses that have been adapted to this diet. The enhanced fatty acid oxidation decreases the production lactic acid during exercise.

There are also specific disease conditions where a higher fat diet is beneficial in preventing signs of the disease such as polysaccharide storage disease and inflammatory bowel disease. Fat supplementation should be excluded in the face of diseases involving liver function or altered lipid metabolism.

Another potential benefit of supplementing lipids is the addition of fatty acids. Though little work has been done with fatty acids in horses, other species have shown great benefit when certain fatty acids levels are increased, especially linoleic and alpha-linolenic. The sources of fat are higher in these fatty acids (omega 3 vs. omega 6) produce less inflammatory mediators. Changing the fatty acid ratios in the diet may be beneficial in reducing the inflammatory response, but more research needs to be done to prove benefits of ongoing supplementation of equine rations.

Reproduction

Several vitamins and minerals are associated with increased reproductive performance in both males and females. As mentioned earlier, the form that minerals are supplied in is critical. All of the minerals listed below have reported toxic effects on reproduction, when exposure is to the inorganic forms, and beneficial effects when exposure is from organic forms.

Biotin is a water soluble vitamin that has been heavily researched in many species with respect to its effect on female reproductive performance parameters and gamete development. The most well documented effects are on sow conception rates and return to estrus post parturition. Many effects of biotin on prenatal development have also been established.

Boron is an ultra-trace mineral that is a good example of a nutrient that is beneficial to the spermatogenic cycle, embryonic, and fetal development when provided in an organic form. However, if the exposure is to an inorganic form, testicular damage and mutagenic effects are common. Boron is involved in the production of many sex hormones. Chromium is also an ultra-trace mineral whose deficiency can cause a decrease in sperm count, but exposure to inorganic forms can cause severe testicular damage, improper testicular development, or neoplasia. Many positive effects on female reproductive efficiencies and lactation improvements have been found.

Folic acid is a water soluble vitamin that plays critical roles in the normal reproduction of cells. A deficiency of folate has been associated with defects in pre-implantation embryos and the neural, skeletal, digestive and urinary tracts of developing fetuses.

Manganese is a trace mineral that is essential for growth, reproduction, prevention of skeletal abnormalities, and congenital ataxia. Mn usually localizes in the cell's mitochrondria. Mn is the metal cofactor (preferred) for a number of glycosyltransferases which provides the link between biochemical function and deficiency symptoms. Mn also plays an important role in carbohydrate, lipid, and brain metabolism. Research has shown that manganese plays a large part in attachment and conception. Exposure to inorganic Mn causes extremely reduced male fertility.

Thiamine is a water soluble vitamin that is crucial to the viability and motility of sperm. Normal development in the uterus depends on the presence of thiamine.

Zinc is a trace mineral whose deficiency has negative effects on testosterone levels and sperm development. Normal growth and lactation are both dependant on the presence of enough zinc. Exposure to inorganic zinc causes harm to male and female reproductive tissues.

AMINO ACIDS AND PROTEIN SUPPLEMENTS

Protein supplementation can be used to support patients with chronic diseases that are in catabolic states. They can also be given when protein is being lost in an abnormal fashion.

If protein supplementation is undertaken, the main selection criteria is biological value of the protein delivered. The main concern with over supplementation of protein or supplementation of lesser quality proteins is the production of toxic waste products such as urea and ammonia. Whey protein is of very high value and is also very economical.

Creatine is an amino acid that provides an energy source for muscle contractions by providing phosphorus to ADP to form ATP. Creatine's contribution occurs during periods of ultra short duration, high exertion. Supplementation of human athletes has shown benefits only in activities such as weightlifting, even sprinting appears to be too long of a duration. Supplementation of horses with creatine has not been shown to improve performance.

Arginine, carnitine and glutamine contribute to glycogenisis. Arginine is also capable of accelerating wound healing and inhibiting the development of some neoplasias. Studies on arginine supplementation have shown significant decreases in recovery time from both trauma and surgeries due to its effects on blood flow, immune function, wound healing, and organ failure.

Carnitine stimulates protein synthesis in the face of stress and may accelerate fatty-acid oxidation, which lowers lactic acid production. Carnitine has also been shown to reduce hepatic fat in several species, via increased production of lipoproteins.

Gastrointestinal

Pre-and Probiotics: Prebiotics are ingredients that when provided to the digestive tract selectively support the growth of beneficial bacterial species over pathogenic ones. Prebiotics do not directly colonize the digestive tract. Prebiotics include yeast, yeast cultures, fungal cultures, and certain fibers (FOS-fructooligosaccharides). Probiotics are the bacterial species that, when introduced to the digestive tract, actually colonize and produce beneficial effects. Synbiotics are products that contain both prebiotic and probiotic ingredients. Ingredients of these types are very important to include in a nutritional support program when the digestive tract is being stressed by changes in ration, environment, location, activity level, etc. Fecal transfer from a healthy donor is a viable way to reintroduce viable microorganisms to the gut environment.

N-acetyl-D-glucosamine: A structural component of all mucosal surfaces. Supplementation with N-acetyl-glucosamine may help firm up the structural matrix of the intestinal tract.

Though glucosamine appears to be highly absorbable, N-acetyl-glucosamine is directly incorporated into the intestinal mucosa and is not absorbed when provided orally. This improves the overall health of the intestinal tract under stress thereby contributing to its healing and increased absorption of other nutrients.

Glutamine supplementation should be incorporated to meet increased energy needs of the enterocytes in diseased or stressed states. 80% of the dietary intake of the amino acid glutamine is used by the enterocytes as energy in normally functioning digestive tracts. Increased intakes will support cell replication and function.

Arginine has been found in human burn patients to promote the release of intestinal hormones and growth factors in the intestinal tract when given orally. It also increases blood flow to the digestive tract.

NUTRITION ECOLOGY

Nutrition ecology is an interdisciplinary scientific discipline that incorporates the entire food chain as well as its interactions with health, the environment, society, and the economy. The food chain includes production, harvesting, preservation, storage, transport, processing, packaging, trade, distribution, preparation, composition, and consumption of food, as well as disposal of all waste materials along the food path.

Nutrition ecology has many roots, some of which go back to antiquity. The introduction of systematic agriculture (slash and burn cultivation) and domestication of animals (food rivals) has markedly affected our environment. One early example of the consequences of systematic agriculture is the Greek invasions of other countries as a consequence of their increasing meat consumption, which required them to acquire more farmland for fodder production. Another example is the deforestation for farmland and for building purposes, which began thousands of years ago and has continued to this day.

Both the Torah and the Bible mention environmental issues numerous times. The impact of systematic agriculture on the environment was discussed by Thomas Aquinas (1224-1274), Jean-Jacques Rousseau (1712-1778), and Henry David Thoreau (1817-1862). At the end of the 19th century, Jacob von Uexkuell (1864-1944) founded the science of ecology.

Industrialized agriculture was introduced in the 19th century and rapidly took command of all aspects of life, with striking social, economic, and environmental consequences. Reactions to these developments led to the formation of the Sierra Club in North America and to the Reform Movement in Central Europe in the second half of the 19th century.

People migrated from urban to rural areas to dwell in unpolluted regions and to grow their own food. Economic and social reforms were proposed and practiced. Some of these included a vegetarian lifestyle.

Another reaction to industrialized agriculture was organic farming, which was initiated by the anthroposophists in 1924 and started to flourish in the 1970s. At that time, a number of organizations were established that raised concerns about the environment and food quality [e.g, the Club of Rome (1968), Greenpeace (1971), World Watch Institute (1975), the Green Party (1980)]. At the same time, literature on the negative influence of industrialized agriculture appeared by Rachel Carson, Frances Moore-Lappe (2), Dennis Meadows, Joan Gussow, and Ralph Nader. These authors discussed the dramatic effects of industrialization and industrialized agriculture on the environment, health, society, and the economy.

ORIGIN OF THE TERM

The term nutrition ecology was coined in 1986 by a group of nutritionists at the University of Giessen, Germany. Nutrition ecology as an interdisciplinary scientific discipline is a holistic concept that considers all links in the nutrition system, with the aim of sustainability. Thus, nutrition ecology describes a new field of nutrition sciences that deals with the local and global consequences of food production, processing, trade, and consumption.

Nutrition ecology goes beyond econutrition, which is limited to the interactions of nutrition and environment. Nutrition ecology goes further than the older concept of ecology of food and nutrition, which is limited to the eating patterns of indigenous and aboriginal populations.

At present, nutrition sciences are dominated by health aspects of food and, in part, by food quality. Recommendations are based primarily on physiologic and toxicologic considerations. The implications of our current nutrition system are more complex and go beyond nutrient content and contamination with pathogens and contaminants. To avoid ecologic damage caused by the nutrition system and to attain nutrition security for the world population, additional aspects need to be incorporated. The necessity of taking a more holistic view for a sustainable development is underlined by the current crises in the nutrition system, as discussed at the World Food Summit in June 2002.

DIMENSIONS OF NUTRITION ECOLOGY

As is typical for an interdisciplinary discipline, nutrition ecology deals with a wide range of issues, including research, teaching, and public actions. A broad view of the entire nutrition system covers subject matters such as total food quality, ecologic balances, and life cycle assessments; the influences of nutrition systems on climate, world nutrition, and food prices; and a comparison of different diets and agricultural, environmental, and consumer policies. Basically, there are four dimensions of nutrition ecology: health, the environment, society, and the economy.

To maintain or retain good health, the consumption of an individually optimal diet is recommended. The term preventative diet has been used recently to underline the possibility of avoiding nutrition-based diseases. The aggregate of most studies suggests that the consumption of plant-derived foods (grains, vegetables, fruits, legumes, nuts) should be increased and that the intake of animal-derived foods (meat products, dairy products, and eggs) should be reduced. This principle applies particularly to sedentary individuals. Plant foods should be consumed when they are as fresh as possible, should be minimally processed, and should be eaten partly as raw food.

The nutrition system influences the environment, which in turn determines the quality of food. The environmental impact of food production is determined by the agricultural method used. Conventional farming methods rely on extensive use of natural resources and result in higher levels of food contamination. In contrast, the environmental impact of organic farming is lower. Organic farming practices include controlling pests naturally, rotating crops, and applying legume plants as manure, in contrast to the use of synthetic pesticides and fertilizers in conventional farming. In integrated farming, organic and conventional methods are combined, resulting in an intermediate environmental impact. To reduce the environmental impact of the nutrition system, organic farming needs to be supported globally. In addition, foods should be minimally processed, packaged, and transported.

The nutrition system is closely related to society, including the responsibility for food purchasing and meal preparation, as well as the social implications of the family meal. Furthermore, the interactions between food consumption habits and lifestyle, as well as the social conditions and the wages of people working in the nutrition system, need to be considered. Additional social aspects include the import and export of agricultural and other products and the influence of this trade on people in developing countries.

On a worldwide basis, the major factor driving food consumption patterns is the financial situation of countries, different population groups, and influential individuals. Transportation and processing of food are carried out under the premise that money can be earned. In private households, the food budget is a determining factor in the choice of foods. From a holistic point of view, the food price should include all costs caused by the nutrition system, especially environmental damage (internalization of external costs). These 4 dimensions of nutrition ecology are of equal importance for achieving a sustainable nutrition system. On this basis, the various aspects of food and nutrition are taken into account. What eating pattern best serves the holistic and sustainable aspects of nutrition ecology? From all we know, a vegetarian diet comes closest to fulfilling the demands and to minimizing damage to the 4 dimensions.

Contribution of Vegetarian Diets

Vegetarians have many reasons not to eat the flesh of animals. In addition to religious beliefs, there are health-based, ecologic, ethical, and philosophical reasons. When the ecologic damage caused by industrial animal production is examined, certain aspects need to be considered. On average, land requirements for meat-protein production are 10 times greater than for plant-protein production. About 40% of the world's grain harvest is fed to animals. Half of this grain would be more than enough to feed all hungry people of our planet. Animal manure, which is produced in huge amounts by industrial agriculture, causes high levels of potentially carcinogenic nitrates in drinking water and vegetables. Animal production requires considerable energy and water resources and leads to deforestation, overgrazing, and overfishing.

One solution to the problems caused by industrial animal production is a vegetarian lifestyle. The positive ecologic effects achieved by vegetarianism can be enhanced by avoiding processed and packaged foods and by choosing seasonally available and locally produced organic foods. In this way, support is given to subsistence and family farming, the securing of employment, and global food security. In addition to these socioeconomic benefits, the caging of animals as well as their transportation over long distances and finally slaughtering them can be avoided, thus fulfilling ethical concerns.

Sustainability

The four dimensions of nutrition ecology are the basis for sustainable nutrition behavior. The term sustainability was introduced in the 17th century by forestry experts in Germany to call attention to the fact that only the amount of trees that would grow back in a given time should be harvested. Presently, sustainability describes development that fulfills current global needs without diminishing the possibility of future generations to meet their own needs.

From a nutritional point of view, sustainability also deals with the fair distribution of food through ecologic and preventive eating behavior. To achieve sustainability, a comprehensive rethinking of common values is needed to attain a new understanding of the quality of life. The question as to the adequate amount of food needs to be addressed at all social levels with the goal of achieving nutrition security for all. To fulfill the demands concerning ecologic, economic, social, and health compatibility, the following 7 principles have been formulated:

(1) food should be predominantly plant derived,

(2) food should originate from organic farming,

(3) food should be produced regionally and seasonally,

(4) food should be minimally processed,

(5) food should be ecologically packaged,

(6) food trade should be fair, and

(7) food should be tastefully prepared.

These principles have been derived from guidelines of wholesome nutrition described elsewhere. A diet based on these principles has a scientific basis, is socially acceptable, is economically feasible, is culturally desired, is practicable, and has a high degree of sustainability.

There are only a limited number of long-term trials on sustainability. In one project, 3 apple production systems-organic farming, integrated farming, and conventional farming-were compared. The yields were nearly equal, but the organic production system showed not only the best apple quality but also the best soil quality and the least detrimental environmental impact. Therefore, the organic production system had the best environmental sustainability. The economic sustainability is given, since the market price was highest for the organic apples.

The authors of this report question the sustainability of conventional farming systems because of escalating production costs, heavy reliance on nonrenewable resources, reduced biodiversity, water contamination, soil erosion, and health risks to farmworkers caused by pesticide use. Another study carried out over 21 y showed that although the crop yield was 20% lower in the organic systems, the input of fertilizer and energy was reduced by 34-53% and the pesticide input by 97%. Enhanced soil fertility and higher biodiversity found in organic plots were due to compost-and legume-based crop rotations.

Biodiversity is also the basis of food variety. Apart from the promotion of breast-feeding, the recommendation to eat a variety of foods is the most internationally agreed-upon dietary guideline. Biodiversity also protects against climate and pestilence disasters. In addition, biodiversity serves increasingly as the basis for new pharmaceuticals.

CONCLUSIONS

Nutrition ecology has the goal of attaining sustainability of food and nutrition security worldwide. To achieve this goal, professionals involved in the nutrition system must inform the public about the principles of nutrition ecology. In this manner, people can be motivated to practice sustainable eating behaviour. Nutrition ecology is also a question of personal priorities. Interested and well-informed consumers will be able to weigh the arguments and make the necessary decisions. The vision of a sustainable future depends upon individuals who feel responsible for the environment and health. One of the most effective ways to achieve the goals of nutrition ecology, including healthy and sustainable food choices, is a vegetarian lifestyle.

SOIL RESEARCH PROGRAMS

Soil science is highly diverse. It covers all aspects of the structure and function of the surface few meters of terrestrial ecosystems. Soil science also interfaces with many other scientific disciplines. At Oregon State University, we have chosen to concentrate on five disciplinary areas within soil science: The following paragraphs give a brief description of each of these areas, provide some examples of current research, and identify faculty who do research in these areas.

Soil Biology

Research in soil biology is centered on the organisms that inhabit the soil, the processes they mediate, and how these organisms and processes are influenced by the soil environment. Soil biology is an important part of current issues such as sustainability of terrestrial ecosystems, soil quality, biodiversity, biodegradation, waste management, and climate change; as well as more traditional issues of nutrient cycling and plant-microbe interactions. Peter Bottomley studies soil microbial ecology, the Rhizobium-legume symbiosis, and xenobiotic degradation.

Richard Dick works on soil enzymes, microbial diversity, soil quality and C sequestration. David Myrold studies nitrogen cycling, the use of stable isotopes in soil science, and the Frankia-actinorhizal plant symbioses. Jennifer Parke studies rhizosphere ecology, biological control of plant diseases, and the soil ecology of plant pathogens and human pathogens. In October, 1999 we established a Microbial Observatory at the H.J. Andrews Experimental Forest with the support of a grant from the National Science Foundation.

Soil Chemistry

Research in soil chemistry focuses on the chemical composition of soils and the chemical reactions that take place at soil surfaces and within the soil solution. Soil chemistry is important for understanding the formation of soils, the fate and reactivity of organic and inorganic pollutants, and availability of plant nutrients, all of which are relevant to current issues of water quality and climate change. John Baham studies the redox reactions of hydric soils, the chemistry of soil formation, and phosphorus availability. Van Volk is interested in using plants as phytoaccumulators of metals, such as nickel, and their application in biomining.

Soil Fertility

Research in soil fertility concentrates on the ability of soils to provide adequate amounts of essential nutrients to plants and environmentally sound management of fertilizers. Research ranges from very basic to applied in nature, and has application to both agronomic production and environmental quality.

Neil Christensen studies interactions between soil fertility and plant pathology, plant nutrition, and nitrogen nutrition of crops. John Hart is an extension specialist who optimizes fertilizer recommendations and studies the fertilizer requirements of a diverse array of crops, such as cranberries, mint, and sweet corn. Dan Sullivan does research on environmentally sound utilization of wastes as soil amendments and sources of plant nutrients.

Soil Landscape Relationships

Research on soil landscape relationships includes the study of pedogenic processes, the description and mapping of soils, and the application of soils information.

Understanding soil formation and applying soils information is vital to current issues such as land use, wetlands delineation, water quality, and regional-to global-scale models of biogeochemical processes. Jay Noller studies soil across landscapes and through time to constrain regional and global scale evolutionary pathways of ecosystems. Herb Huddleston is an extension specialist who studies hydric soils and use of soils information in GIS applications. Implications of Long-Term Land Use and Land Cover Change on Cyprus: A Holocene History of Soil Erosion and Conservation

Soil Physics

Research on soil physics emphasizes the physical properties of soil and the transport of energy, water, solutes, and gases. Soil physics is important to proper soil and agronomic management, related to issues of tillage, erosion, irrigation, and drainage; it also plays a central role in environmental problems, such as water quality and climate change. Larry Boersma does research on plant-water relations, water transport, and modeling soil physical processes.

Rich Roseberg is located at Medford and studies water use. Benno Warkentin does research on groundwater quality. Don Wysocki is located at Pendleton and studies soil management in dryland wheat systems. Maria Dragila does research in water transport, preferential flow and pore scale processes.

BASIC NUTRITION

Many people worry that when they stop eating meat and fish, they might be in danger of some nutritional deficiency. This is not the case as all the nutrients you need can easily be obtained from a vegetarian diet. In fact research shows that in many ways a vegetarian diet is healthier than that of a typical meat-eater.

Nutrients are usually divided into five classes: carbohydrates, proteins, fats (including oil), vitamins and minerals. We also need fibre and water. All are equally important to our well-being, although they are needed in varying quantities, from about 250g of carbohydrate a day to less than two micrograms of vitamin B12. Carbohydrate, fat and protein are usually called macro-nutrients and the vitamins and minerals are usually called micro-nutrients.

Most foods contain a mixture of nutrients (there are a few exceptions, like pure salt or sugar) but it is convenient to classify them by the main nutrient they provide. Still, it is worth remembering that everything you eat gives you a whole range of essential nutrients. Meat supplies protein, fat, some B vitamins and minerals (mostly iron, zinc, potassium and phosphorous). Fish, in addition to the above, supplies vitamins A, D, and E, and the mineral iodine. All these nutrients can be easily obtained by vegetarians from other sources, as this Information Sheet shows.

PROTEIN

Women need about 45g of protein a day (more if pregnant, lactating or very active), men need about 55g (more if very active). Evidence suggests that excess protein contributes to degenerative diseases. Vegetarians obtain protein from:

- Nuts: hazels, brazils, almonds, cashews, walnuts, pine kernels etc.
- Seeds: sesame, pumpkin, sunflower, linseeds.
- Pulses: peas, beans, lentils, peanuts.
- Grains/cereals: wheat (in bread, flour, pasta etc), barley, rye, oats, millet, maize (sweetcorn), rice.

- Soya products: tofu, tempeh, textured vegetable protein, veggieburgers, soya milk.
- Dairy products: milk, cheese, yoghurt (butter and cream are very poor sources of protein).
- Free range eggs.

You have may have heard that it is necessary to balance the complementary amino acids in a vegetarian diet. This is not as alarming as it sounds. Amino acids are the units from which proteins are made. There are 20 different ones in all. We can make many of them in our bodies by converting other amino acids, but eight cannot be made, they have to be provided in the diet and so they are called essential amino acids. Single plant foods do not contain all the essential amino acids we need in the right proportions, but when we mix plant foods together, any deficiency in one is cancelled out by any excess in the other. We mix protein foods all the time, whether we are meat-eaters or vegetarians. It is a normal part of the human way of eating. A few examples are beans on toast, muesli, or rice and peas. Adding dairy products or eggs also adds the missing amino acids, eg macaroni cheese, quiche, porridge. It is now known that the body has a pool of amino acids so that if one meal is deficient, it can be made up from the body's own stores. Because of this, we don't have to worry about complementing amino acids all the time, as long as our diet is generally varied and well-balanced. Even those foods not considered high in protein are adding some amino acids to this pool.

CARBOHYDRATE

Carbohydrate is our main and most important source of energy, and most of it is provided by plant foods. There are three main types: simple sugars, complex carbohydrates or starches and dietary fibre.

The sugars or simple carbohydrates can be found in fruit, milk and ordinary table sugar. Refined sources of sugar are best avoided as they provide energy without any associated fibre, vitamins or minerals and they are also the main cause of dental decay. Complex carbohydrates are found in cereals/grains (bread, rice, pasta, oats, barley, millet, buckwheat, rye) and some root vegetables, such as potatoes and parsnips. A healthy diet should contain plenty of these starchy foods as a high intake of complex carbohydrate is now known to benefit health.

The unrefined carbohydrates, like wholemeal bread and brown rice are best of all because they contain essential dietary fibre and B vitamins. The World Health Organisation recommends that 50-70% of energy should come from complex carbohydrates. The exact amount of carbohydrate that you need depends upon your appetite and also your level of activity. Contrary to previous belief a slimming diet should not be low in carbohydrates. In fact starchy foods are very filling relative to the number of calories that they contain.

DIETARY FIBRE

Dietary fibre or non-starch polysaccharide (NSP), as it is now termed, refers to the indigestible part of a carbohydrate food. Fibre can be found in unrefined or wholegrain cereals, fruit (fresh and dried) and vegetables. A good intake of dietary fibre can prevent many digestive problems and protect against diseases like colon cancer and diverticular disease.

FATS AND OILS

Too much fat is bad for us, but a little is necessary to keep our tissues in good repair, for the manufacture of hormones and to act as a carrier for some vitamins. Like proteins, fats are made of smaller units, called fatty acids. Two of these fatty acids, linoleic and linolenic acids, are termed essential as they must be provided in the diet. This is no problem as they are widely found in plant foods. Fats can be either saturated or unsaturated (mono-unsaturated or poly-unsaturated).

A high intake of saturated fat can lead to a raised blood cholesterol level and this has been linked to heart disease. Vegetable fats tend to be more unsaturated and this is one of the benefits of a vegetarian diet. Mono-unsaturated fats, such as olive oil or peanut oil, are best used for frying as the poly-unsaturated fats, like sunflower or safflower oil are unstable at high temperatures. Animal fats (including butter and cheese) tend to be more saturated than vegetable fats, with the exception of palm oil and coconut oil.

VITAMINS

Vitamin is the name for several unrelated nutrients that the body cannot synthesise either at all, or in sufficient quantities. The one thing they have in common is that only small quantities are needed in the diet. The main vegetarian sources are listed below:

Vitamin A (or beta carotene): Red, orange or yellow vegetables like carrots and tomatoes, leafy green vegetables and fruits like apricots and peaches. It is added to most margarines.

B Vitamins: This group of vitamins includes B1 (thiamin), B2 (riboflavin), B3 (niacin), B6 (pyridoxine), B12 (cyanocobalmin), folate, pantothenic acid and biotin.

All the B vitamins except B12 occur in yeasts and whole cereals (especially wheat germ), nuts & seeds, pulses and green vegetables.

Vitamin B12 is the only one that may cause some difficulty as it is not present in plant foods. Only very tiny amounts of B12 are needed and vegetarians usually get this from dairy produce and free range eggs. It is sensible for vegans and vegetarians who consume few animal foods to incorporate some B12 fortified foods in their diet. Vitamin B12 is added to yeast extracts, soya milks, veggieburgers and some breakfast cereals.

***Vitamin** C:* Fresh fruit, salad vegetables, all leafy green vegetables and potatoes.

***Vitamin** D:* This is vitamin is not found in plant foods but humans can make their own when skin is exposed to sunlight. It is also added to most margarines and is present in milk, cheese and butter. These sources are usually adequate for healthy adults. The very young, the very old and anyone confined indoors would be wise to take a vitamin D supplement especially if they consume very few dairy products.

Vitamin E: Vegetable oil, wholegrain cereals, eggs.

Vitamin K: Fresh vegetables, cereals and bacterial synthesis in the intestine.

MINERALS

Minerals perform a variety of jobs in the body. Details of the some of the most important minerals are listed below:

Calcium: Important for healthy bones and teeth. Found in dairy produce, leafy green vegetables, bread, tap water in hard water areas, nuts and seeds (especially sesame seeds), dried fruits, cheese. Vitamin D helps calcium to be absorbed.

Iron: Needed for red blood cells. Found in leafy green vegetables, wholemeal bread, molasses, eggs, dried fruits (especially apricots and figs), lentils and pulses. Vegetable sources of iron are not as easily absorbed as animal sources, but a good intake of vitamin C will enhance absorption.

Zinc: Plays a major role in many enzyme reactions and the immune system. Found in green vegetables, cheese, sesame and pumpkin seeds, lentils and wholegrain cereals.

Iodine: Present in vegetables, but the quantity depends on how rich the soil is in iodine. Dairy products also have plenty of iodine. Sea vegetables are a good source of iodine for vegans.

CALCIUM

Calcium is the most abundant mineral in the human body. Of the body's total calcium, about 99% is in the bones and teeth where it plays a structural role. The remaining 1% is present in body tissues and fluids where it is essential for cell metabolism, muscle contraction and nerve impulse transmission.

Functions

The main function of calcium is structural. The skeleton of a young adult male contains about 1.2 kg of calcium. There is continuous movement of calcium between the skeleton and blood and other parts of the body. This is finely controlled by hormones. Metabolites of Vitamin D are important in this, increasing reabsorption of calcium by bones. Calcium also plays a role in cell biology. Calcium can bind to a wide range of proteins altering their biological activity. This is important in nerve impulse transmission and muscle contraction. Calcium is also needed for blood clotting, activating clotting factors. Vitamin D is needed for absorption of dietary calcium and so calcium deficiency may be linked with rickets in children. In adults, calcium deficiency may lead to osteomalacia (softening of bones). This may be related to repeated pregnancy with lengthy breast feeding.

Osteoporosis can be due to calcium deficiency. This involves loss of calcium from the bones and reduced bone density. This causes bones to be brittle and liable to fracture. Bone loss occurs with age in all individuals. This usually occurs after 35-40 years and involves the shrinking of the skeleton. Bone loss is greatest in women following the menopause. This is due to reduced levels of the hormone, oestrogen. Postmenopausal women are particularly at risk from osteoporosis.

Some research has indicated vegetarian women are at less risk of osteoporosis than omnivorous women. This is thought to be due to animal protein increasing calcium loss from bones. However, other research has found no difference between vegetarians and omnivores. The risk of osteoporosis may be altered by factors other than diet. Lack of exercise, being underweight, smoking and alcohol can all increase the risk. A low level of calcium in the blood and tissues can cause hypocalcaemia.

This involves sensations of tingling and numbness and muscle twitches. In severe cases muscle spasms may occur. This is called tetany. It is more likely to be due to a hormonal imbalance in the regulation of calcium rather than a dietary deficiency. Excess calcium in the blood can cause nausea, vomiting and calcium deposition in the heart and kidneys. This usually results from excessive doses of vitamin D and can be fatal in infants.

DIETARY SOURCES

Calcium is present in a wide range of foods. Dairy products, leafy green vegetables, nuts and seeds (almonds, brazils, sesame seeds), tofu, and dried fruit are all good sources of calcium for vegetarians.

Most flour is fortified with calcium carbonate so cereals can also be a good source. Hard water may also provide calcium. Meat is a very poor source of calcium. Calcium balance can be affected by a range of other factors. Vitamin D is essential for absorption of calcium from the gut. This is because calcium is transported into the body by a special carrier protein which requires vitamin D for its synthesis.

Sources of Calcium (single servings)					
Good Sources		Fair Sources		Poor sources	
Tofu (60g or 2oz)	304 mg	Brown bread (2 slices)	70 mg	Spaghetti, boiled (100g or 3½oz)	7 mg
Cheddar cheese (slice, 40g)	288 mg	Brazil nuts (9 nuts, 30g)	54 mg	Brown rice (190g or 8oz)	7 mg
Cows milk (0.3 pint)	234 mg	Dried apricots (8 apricots)	46 mg	-	-
Spinach, boiled (130g or 5oz)	208 mg	French beans, boiled (100g)	41 mg	-	-
Dried figs (4 figs)	168 mg	Cottage cheese (45g or 1½oz)	33 mg	-	-
Soya cheese (slice, 40g)	180 mg	Sesame seeds (15g or ½oz)	20 mg	-	-
Chick peas, boiled (200g or 8oz)	92 mg	-	-	-	-
Baked beans (200g or 8 oz)	90 mg	-	-	-	-
Broccoli, boiled (95g or 3½oz)	72 mg	-	-	-	-

A number of substances can inhibit the absorption of calcium. Phytic acid, found in bran, whole cereals and raw vegetables is one of these. Uronic acid, a component of dietary fibre, and oxalic acid, found in certain fruits and vegetables can also bind calcium.

However, diets habitually high in these acids are not thought to have a major effect on calcium absorption. Saturated fats can also lessen calcium absorption. Calcium is lost in the faeces, urine and sweat. Calcium loss is roughly equal to dietary calcium in adults. Calcium loss is reduced if dietary calcium is low. Adaptation to both high and low calcium intakes occur. Reduced intake leads to increased efficiency of absorption. In infants and children calcium is retained for new bone growth. Calcium is also lost during lactation in breast milk.

Required Intakes

The old Recommended Daily Amounts (RDAs) have now been replaced by the term Reference Nutrient Intake (RNI). The RNI is the amount of nutrient which is enough for at least 97% of the population.

Nutrient intakes for calcium, mg/day.

Age	RNI	Age	RNI
0 to 12 months	525 mg	11 to 18 years, male	1000 mg
1 to 3 years	350 mg	11 to 18 years, female	800 mg
4 to 6 years	450 mg	19 + years	700 mg
7 to 10 years	550 mg	Breast feeding women	1250 mg

During pregnancy, calcium absorption from the gut increases and no additional calcium is generally needed. Pregnant adolescents are an exception to this, having particularly high calcium needs. Breast feeding women need an extra 550 mg of calcium. A lactating women can lose up to 300 mg a calcium/day in breast milk. Calcium absorption decreases with age so it is important the elderly have adequate dietary calcium.

Meal Plan

Sample one day's meal plan to meet the RNI of 700 mg for an adult					
Breakfast		**Lunch**		**Evening meal**	
Muesli	114 mg	2 slices of brown bread toast	140 mg	Tofu (60g) with broccoli (95g)	376 mg
Soya milk, unfortified	15 mg	Baked beans (200g or 8oz)	90 mg	Brown rice, boiled (165g)	7 mg
Total calcium intake - 742 mg					

FATS AND CHOLESTEROL

Fats provide a concentrated source of energy in the diet. The building blocks of fats are called fatty acids. These can be either saturated, monounsaturated or poly-unsaturated. Foods rich in saturated fats are usually of animal origin. Vegetable fats are generally unsaturated. Saturated fat raises the level of cholesterol in the blood. Cholesterol is present in animal foods but not plant foods. It is essential for metabolism but is not needed in the diet as our bodies can produce all that is needed. Raised blood cholesterol is associated with an increased risk of heart disease. Fats and oils are essentially the same. Fats tend to be solid at room temperature whilst oils are liquid. The term lipids include both fats and oils.

Structure and Functions

Fats consist of fatty acids and glycerol. Nearly all the fats in our bodies and in foods are triglycerides, being made up of three fatty acid molecules to one glycerol molecule. There are about 16 different fatty acids commonly present in foods. The nature of fat depends on its constituent fatty acids. Fats can be classed as either saturated, monounsaturated or polyunsaturated. This depends on the type of chemical bonds present in the fatty acid.

If a fatty acid has all the hydrogen atoms it can hold it is termed saturated. However, if some of the hydrogen atoms are absent and the usual single bond between carbon atoms has been replaced by a double bond, then it is unsaturated. If there is just one double bond then it is monounsaturated. If there is more than one then it is polyunsaturated. Most fats contain a proportion of each of these three basic types of fatty acid but are generally described according to which type predominates.

Saturated fats tend to be animal fats and are solid at room temperature. Butter, lard, suet and meat fat are saturated fats. Unsaturated fats are liquid at room temperature. They are usually of plant origin, though fish oils may also be high in polyunsaturated fatty acids. Plant oils may be hardened by the addition of hydrogen atoms, converting double bonds to single bonds. This process is known as hydrogenation. Hydrogenated vegetable oils are often present in margarine and other processed foods. Fats have a number of important functions in the body. As well as being a concentrated source of energy, fats act as carriers for fat-soluble vitamins A, D, E and K. Fats are also essential for the structure of cell membranes and are precursors of many hormones.

Essential Fatty Acids

Two fatty acids are termed essential fatty acids. These are linoleic acid and a-linolenic acid. These must be present in the diet as the body is unable to make them itself. They are widely present in plant oils such as sunflower, rapeseed and soyabean oils.

Linoleic acid is converted into the body to arachidonic acid from which prostoglandins and other vital compounds are made. Because of this conversion, arachidonic acid is not an essential fatty acid as was once believed. a-Linolenic acid is converted to eicosapentanoic acid (EPA) which is important in proper nerve function.

EPA is present in fish oils and is claimed to be beneficial in reducing the symptoms of arthritis and the risk of heart disease. For this reason, fish oils are sometimes used therapeutically. Plant oils containing large amounts of a-linolenic acid can be used as an alternative by vegetarians. Linseeds and linseed oil are particularly rich sources of a-linolenic acid.

Cholesterol

Cholesterol belongs to the sterol group of fats. It is present in all animal tissues but is absent from plants. Cholesterol is essential as a component of cell membranes and a precursor of bile acids and certain hormones.

The body can make its own cholesterol and so a dietary source is not required. Cholesterol is transported in to various proteins. These complex molecules are called lipoproteins. There are four main types of lipoprotein involved in cholesterol transport. The most commonly refer red to are low density lipoprotein (LDL) and high density lipoprotein (HDL).

Cholesterol may form plaques on artery walls if levels in the blood are too high. This can lead to atherosclerosis. Because of this high blood cholesterol is linked with heart disease. It is the LDL cholesterol which has been linked to heart disease.

HDL cholesterol may help protect against the risk of heart disease. The amount of dietary cholesterol is not clearly linked to levels of cholesterol in the blood. Blood cholesterol is more closely related to the amount of saturated fat in the diet, saturated fat raising blood cholesterol. Unsaturated fats are not thought to raise blood cholesterol and may indeed lower levels.

Trans Fatty Acids

Unsaturated fatty acids can exist in two different geometric forms. These are called the cis and trans forms. Unsaturated fatty acids exist naturally in the cis form. During food manufacturing processes these cis fatty acids may be changed to trans fatty acids. Hydrogenation of margarine causes this to occur. It has been suggested that trans fatty acids can increase the risk of heart disease.

Free Radicals

Free radicals are highly reactive molecules which have been linked to both heart disease and cancer. A number of factors, including alcohol, stress and environmental pollutants can increase the generation of free radicals in the body. Polyunsaturated fats can also generate free radicals, especially when exposed to heat or sunlight. Because of this it is suggested that vegetable oils should be stored out of direct sunlight. Mono-unsaturated olive oil is less vulnerable to free radical generation and so is a better choice for frying. Anti-oxidants such as vitamins A, C and E offer protection against free radicals. Fresh fruit and vegetables are rich in these anti-oxidants.

Dietary Sources

Saturated fats are nearly always from animal foods. Meat, eggs and dairy products all contain saturated fats. Lard and suet are saturated fats. Coconut oil and palm oil are vegetable sources of saturated fats. Olive oil is a monounsaturated fat. Polyunsaturated fats are usually from plant sources. The ratio of polyunsaturated to saturated fats in the diet is often called the P:S ratio. Cholesterol is present in all animal foods but not plant foods. Egg yolks and high-fat dairy products are high in cholesterol.

Required Intakes

Currently it is believed that around 42% of energy in the typical British diet is from fat. Dietary advice is to reduce this. The COMA (Committee on Medical Aspects of Food Policy) report advocated that no more than 35% of daily energy requirement should come from fat whilst the NACNE (National Advisory Committee on Nutritional Education) paper recommends a reduction to no more than 30%. Special emphasis is placed on reducing the amount of saturated fat in the diet. Vegetarian diets tend to be lower in fat than omnivore diets.

However, vegetarians consuming dairy products and processed foods high in fat may still be consuming too much. Advice to vegetarians is to keep fat intake to a minimum, avoid high fat dairy products and processed foods containing dairy fats and hydrogenated vegetable fats, and to use olive oil for cooking purposes.

CEREALS

All too often, cereal products are thought of as nothing more than starchy fillers, and indeed, when you consider how some commercial products like cakes and biscuits and breakfast cereals are made from over-refined grains with nearly all the nutrients except the starch extracted, and then loaded with fat, sugar, artificial flavourings etc., you might be forgiven for thinking that's true. But go back to the original, unrefined grains and you have a wealth of nutrients in a small package. Grains have been the staple foods of many civilisations for thousands of years. Wheat, barley, oats and rye in Europe, maize in America, quinoa in South America, rice in the East, and millet in Africa.

Nutrition: Cereals are seeds of plants, usually members of the grass family but there are a few exceptions. They are annuals, that is they have to be planted every year and at the end of the summer, when they have produced ripe seeds, they die down. Like all seeds, cereals are very nutritious because they contain all the nutrients the embryo plant needs to start growing.

Unrefined cereals are valuable sources of proteins, carbohydrates, B vitamins and also contain some fat, iron, vitamin E and trace minerals and are a very good source of fibre in the diet. Some cereals (wheat, barley, rye and oats) contain the protein gluten, which is essential for leavened bread-making. Without sufficient gluten, bread will not rise. People suffering from gluten intolerance or coeliac disease must avoid any cereal containing gluten. In some parts of the world unleavened (unrisen) bread is eaten r the staple cereal is made into noodles or pasta. They are usually cheap to buy and are a valuable source of variety in the vegetarian diet.

Storage: Keep in airtight containers in a cool, dark, dry place. Whole grains can be stored for up to 2 years; flaked, cracked grains and flours should be used within 2-3 months of purchase.

Sprouting: Whole grains can be sprouted, which greatly enhances their nutritional value, e.g. wheat grains, raw buckwheat and barley.

Cooking: Cereals can be used in other ways, besides being ground into flour for bread, cakes etc. Whole grains can be added to stews and casseroles, or cooked until soft. Cracked or kibbled grains are cut or broken pieces of whole grains e.g. kibbled wheat and bulgur wheat. Meal, a coarse kind of flour, can be used to make porridge, thicken soups or mixed with wheat flour to add interesting flavours and textures to ordinary breads, biscuits, muffins etc. Whole grains should be washed thoroughly. Boil the required amount of water, add the washed grain, stir once, put a tight-fitting lid on the pan and simmer for the required cooking time or until the liquid is absorbed. Turn off the heat and leave to stand for 5 minutes before removing the lid.

Barley

Barley grows in a wider variety of climatic conditions than any other cereal. It used to be a very important source of direct human food, but its use has diminished over the last 250 years, replaced by wheat, and it is now used almost exclusively as animal feed or for making beer and whisky. It contains gluten, so barley flour can be made into bread. More usually found in the shops as whole or pot barley, or polished pearl barley, it is also possible to buy barley flakes or kernels. The whole barley is more nutritious with 100g providing 10.5g protein, 2.1g fat, 69.3g carbohydrate, 4g fibre, 50mg folic acid, 6mg iron and 50mg calcium. It can be cooked on its own (1:3 parts water for 45-60 minutes) as a pleasant alternative to rice, pasta or potatoes, or added to stews. Malt extract is made from sprouted barley grains.

Buckwheat (Gluten Free)

Buckwheat is not a true cereal as it is not a member of the grass family, instead being related to sorrels and docks. If you look at docks closely, you can see that the seeds, though smaller, have the same distinctive triangular shape. Buckwheat, a native of central Asia, is now grown in Europe, N America and the former USSR countries, but it is still not widely used in Britain. 100g of buckwheat provides 11.7g protein, 3.9mg iron and it is very high in calcium with 114mg per 100g. Available raw the seeds are greenish-pink, or roasted (known as kasha) the seeds are darker reddish-brown. It can be cooked (1:2 parts water for 6 minutes, leave to stand for 6 minutes) and served like rice or you can add it to stews and casseroles. Buckwheat flour can be added to cakes, muffins, pancakes etc. where it imparts a distinctive flavour. Look out too for buckwheat spaghetti, soba.

Corn or Maize (Gluten Free)

Maize is the principal food plant of America and was unknown in other parts of the world until Columbus reached America in 1492. It was grown by the Maya, Inca and Aztec civilisations, and by various North American Indian tribes and now has spread to Canada, USSR, Italy, Spain, Egypt, India and South Africa. It is used for human food, animal feed and as a source of raw materials for industry. 100g maize gives 9g protein. Fresh maize is often available (sweet corn, corn on the cob) but remember that nutritional values will be lower because less concentrated. We most often see maize as cornflakes or popcorn, but cornmeal or polenta is available and can be added to soup, pancakes, muffins etc. Tortillas are made from maize meal, as are quite a lot of snack foods. Do not confuse cornmeal with highly refined corn starch/ flour, used for thickening.

Millet (Gluten Free)

Millet is the name applied to a variety of grasses first cultivated in Asia or Africa. It is a staple crop in Africa because it is drought resistant and keeps well. 100g millet provides 9.9g protein, and 6.8mg iron (higher than other cereals). Millet makes a delicious alternative to rice but the tiny seeds need to be cracked before they will absorb water easily, so they should be first sauteed with a little vegetable oil for 2-3 minutes until some are seen to crack, then add water with care (1:3 parts), bring to the boil and simmer for 15-20 minutes until fluffy. Millet flakes can be made into porridge or added to muesli and millet flour is available, sometimes also made into pasta.

Oats

Oats are thought to have originated in Western Europe and may originally have appeared as a weed in barley and so got spread with the barley. They're now grown in many parts of the world including N W Europe, the former USSR countries, North America, Canada, Australia and China. Used mostly as animal feed, they are very nutritious. In fact, as they are usually inexpensive to buy, they can be a real boon to people trying to get a good diet on a low budget. 100g oats gives 13g protein, 55mg calcium (more than any other cereal except buckwheat), and 4.6mg iron. Available as groats (whole grains with the husks removed) but more usually as various grades of oatmeal, rolled oats or jumbo oat flakes. Oat groats need cooking for 45 minutes in 1:3 parts water. All forms can be used to make porridge, combined with ground nuts to make a roast or added to stews. Oatmeal is low in gluten so can't be used to make a loaf, but can be mixed with wheat flour to add flavour and texture to bread, muffins, pancakes etc.

Quinoa (Gluten Free)

Quinoa is an ancient crop which fed the Central American Aztec Indians for thousands of years, and which has recently been cultivated in Britain. Unlike most grains, it does not belong to the grass family, but is a relative of the garden weed called Fat Hen. It is very nutritious, containing between 13-14% protein with a good amino acid composition but has no gluten so it cannot be used for bread-making. Instead, it is cooked for 15 minutes in 1:3 parts water and served as a side dish or may be used in risotto, pilaff, vegetable stuffings etc.

Rice (Gluten Free)

Rice is one of the world's most important crops. It originated in Asia but is now grown throughout the humid, sub-tropical regions. It differs from most other cereals in requiring land that is submerged in water to grow, though some varieties do grow in upland areas. Rice is a good source of carbohydrate but doesn't have quite as much protein as some other cereals (6.5g per 100g).

Unpolished rice (i.e. wholegrain/brown rice) is a good source of B vitamins too. There are three basic kinds in culinary terms: long, medium and short grain. Long used traditionally in savoury dishes, short in dessert cooking, although this varies across the globe and it is really a matter of personal preference. Long wholegrain rice needs to cook in 1:2 parts water for 35-40 minutes. Rice flour is available but because of the lack of gluten, it cannot be used to make a yeasted loaf but can be used for cakes, biscuits and pancakes. Rice flakes (brown and white) can be added to muesli or made into a milk pudding or porridge.

Wild Rice

Wild rice is not a rice at all but an American grass used as an important food by the Indians and early settlers. Difficulty in harvesting makes it expensive, but the colour, a purplish black and its subtly nutty flavour make it a good base for a special dish and it can be economically mixed with other rices, but may need pre-cooking as it takes 45-50 minutes to cook in 1:3 parts of water.

Rye

Rye is the least important cereal crop and is usually only grown where conditions are relatively unfavourable and other cereals don't do well. It probably originated in S W Asia, but the name occurs in Northern European languages, which suggest early cultivation in that area. It is very hardy and so grows in temperate and cool regions and at high altitudes, and is very tolerant of poor soil fertility.

It is the only cereal apart from wheat and barley that has enough gluten to make a yeasted loaf, but it has less gluten than wheat, so rye bread is denser. It is more usual to mix rye flour with wheat flour. Rye grains should be cooked in 1:3 parts water for 45-60 minutes. Kibbled rye is often added to granary-type loaves. You can also add rye to stews and rye flakes are available, which can be used in muesli. 100g of rye gives 9.4g protein.

Spelt

Spelt is closely related to common wheat, originating in the Middle East, and has been popular for decades in Eastern Europe. Higher in protein than wheat, it appears to have a different molecular structure, appearing to cause less problems than wheat for some sufferers of grain allergies. It has an intense nutty, wheaty flavour. The flour is excellent for breadmaking and spelt pasta is becoming more widely available.

Wheat

This is the most familiar cereal used in Britain today, it is used for bread, cakes, biscuits, pastry, breakfast cereals and pasta. All the present varieties of wheat seem to be derived from a hybrid wild wheat that grew in the Middle East 10,000 years ago. Over 30,000 varieties are said to be in cultivation. Wheat can be grown in a very wide range of climatic conditions but is most successful in temperate zones including the UK, North America, Southern Russia and South West Australia.

Nutritionally, 100g whole wheat provide 14g protein, 2.2g fat, 69.1g carbohydrate, 2.3g fibre, 3.1mg iron, 36mg calcium. Wheat grains, also called wheat berries, can be eaten whole, cooked in 1:3 parts of water for 40-60 minutes, they have a satisfying, chewy texture. Cracked or kibbled wheat is the dried whole grains cut by steel blades. Bulgur wheat, made from the whole grains steamed before cracking, only needs rehydrating by soaking in boiling water or stock. Couscous is the steamed, dried and cracked grains of durum wheat and is more refined than bulgur.

Soak in 2 parts of water/stock to rehydrate, traditionally it is steamed after soaking. Strong wheat flour (high gluten content) is required for yeasted breadmaking and puff pastry. Plain flour is used for general cooking including cakes and shortcrust pastry. Wheat flakes are used for porridge, muesli and flapjacks. Wheat germ is an excellent source of nutrients, especially vitamin E.

2

PRESERVING THE FOOD: BIOTECHNOLOGICAL METHODS

HOW FOOD PRESERVATION WORKS

Because food is so important to survival, food preservation is one of the oldest technologies used by human beings. In this article, we'll look at all of the different preservation techniques commonly used today, including:

- Refrigeration and freezing
- Canning
- Irradiation
- Dehydration
- Freeze-drying
- Salting
- Pickling
- Pasteurizing
- Fermentation
- Carbonation
- Cheese-making
- Chemical preservation

The basic idea behind all forms of food preservation is either:

- To slow down the activity of disease-causing bacteria
- To kill the bacteria altogether

In certain cases, a preservation technique may also destroy enzymes naturally found in a food that cause it to spoil or discolor quickly. An enzyme is a special protein that acts as a catalyst for a chemical reaction, and enzymes are fairly fragile. By increasing the temperature of food to about 150 degrees Fahrenheit (66 degrees Celsius), enzymes are destroyed.

A food that is sterile contains no bacteria. Unless sterilized and sealed, all food contains bacteria. For example, bacteria naturally living in milk will spoil the milk in two or three hours if the milk is left out on the kitchen counter at room temperature.

By putting the milk in the refrigerator you don't eliminate the bacteria already there, but you do slow down the bacteria enough that the milk will stay fresh for a week or two.

Let's look at all of the different forms of food preservation in detail.

REFRIGERATION AND FREEZING

Refrigeration and freezing are probably the most popular forms of food preservation in use today.

In the case of refrigeration, the idea is to slow bacterial action to a crawl so that it takes food much longer (perhaps a week or two, rather than half a day) to spoil. In the case of freezing, the idea is to stop bacterial action altogether. Frozen bacteria are completely inactive. Refrigeration and freezing are used on almost all foods: meats, fruits, vegetables, beverages, etc.

In general, refrigeration has no effect on a food's taste or texture. Freezing has no effect on the taste or texture of most meats, has minimal effects on vegetables, but often completely changes fruits (which become mushy). Refrigeration's minimal effects account for its wide popularity.

CANNING

Since 1825 or so, canning has provided a way for people to store foods for extremely long periods of time. In canning, you boil the food in the can to kill all the bacteria and seal the can (either before or while the food is boiling) to prevent any new bacteria from getting in. Since the food in the can is completely sterile, it does not spoil. Once you open the can, bacteria enter and begin attacking the food, so you have to "refrigerate the contents after opening" (you see that label on all sorts of food products--it means that the contents are sterile until you open the container).

We generally think of "cans" as being metal, but any sealable container can serve as a can. Glass jars, for example, can be boiled and sealed. So can foil or plastic pouches and boxes. Milk in a box that you can store on the shelf is "canned" milk. The milk inside the box is made sterile (using ultra high temperature (UHT) pasteurization) and sealed inside the box, so it does not spoil even at room temperature.

One problem with canning, and the reason why refrigeration or freezing is preferred to canning, is that the act of boiling food in the can generally changes its taste and texture (as well as its nutritional content).

DEHYDRATION

Many foods are dehydrated to preserve them. If you walk through any grocery store you may notice the following dehydrated products:

- Powdered milk
- Dehydrated potatoes in a box
- Dried fruits and vegetables
- Dried meats (like beef jerky)
- Powdered soups and sauces
- Pasta
- Instant rice.

Since most bacteria die or become completely inactive when dried, dried foods kept in air-tight containers can last quite a long time.

Normally, drying completely alters the taste and texture of the food, but in many cases a completely new food is created that people like just as much as the original!

FREEZE-DRYING

Freeze-drying is a special form of drying that removes all moisture and tends to have less of an effect on a food's taste than normal dehydration does.

In freeze-drying, food is frozen and placed in a strong vacuum. The water in the food then sublimates--that is, it turns straight from ice into vapor. Freeze-drying is most commonly used to make instant coffee, but also works extremely well on fruits such as apples.

An Experiment in Freeze-Drying

You probably don't have a good vacuum chamber at home, but you almost certainly have a refrigerator. If you don't mind waiting a week, you can experiment with freeze-drying at home using your freezer.

For this experiment you will need a tray, preferably one that is perforated. If you have something like a cake-cooling rack or a metal mesh tray, that is perfect. You can use a cookie sheet or a plate if that is all that you have, but the experiment will take longer.

Now you will need something to freeze-dry. Three good candidates are apples, potatoes and carrots (apples have the advantage that they taste okay in their freeze-dried state). With a knife, cut your apple, potato and/or carrot as thin as you can (try all three if you have them).

Cut them paper-thin if you can do it--the thinner you cut, the less time the experiment will take. Then arrange your slices on your rack or tray and put them in the freezer. You want to do this fairly quickly or else your potato and/or apple slices will discolor. In half an hour, look in on your experiment. The slices should be frozen solid.

Over the next week, look in on your slices. What will happen is that the water in the slices will sublimate away. That is, the water in the slices will convert straight from solid water to water vapor, never going through the liquid state (this is the same thing that mothballs do, going straight from a solid to a gaseous state).

After a week or so (depending on how cold your freezer is and how thick the slices are), your slices will be completely dry. To test apple or potato slices for complete drying, take one slice out and let it thaw. It will turn black almost immediately if it is not completely dry.

When all of the slices are completely dry, what you have is freeze-dried apples, potatoes and carrots. You can "reconstitute" them by putting the slices in a cup or bowl and adding a little boiling water (or add cold water and microwave). You can eat the apples in their dried state or you can reconstitute them. What you will notice is that the reconstituted vegetables look and taste pretty much like the original! That is why freeze-drying is a popular preservation technique.

SALTING, PICKLING

Salting, especially of meat, is an ancient preservation technique. The salt draws out moisture and creates an environment inhospitable to bacteria. If salted in cold weather (so that the meat does not spoil while the salt has time to take effect), salted meat can last for years.

The following passage from John Steinbeck's "The Grapes of Wrath" describes the process briefly:

Noah carried the slabs of meat into the kitchen and cut it into small salting blocks, and Ma patted the course salt in, laid it piece by piece in the kegs, careful that no two pieces touched each other. She laid the slabs like bricks, and pounded salt in the spaces.

This technique creates a keg (a wooden barrel) full of salt and meat. This technique is ancient. You can read about its use during the sailing voyages around the time of Columbus. Many accounts of the Revolutionary War and especially the Civil War talk about meat preserved in this way. Salting was used to preserve meat up through the middle of this century, and was eventually replaced by refrigeration and freezing.

Today, salting is still used to create salt-cured "country ham" found widely in the southern United States, dried beef (which you can buy in jars at most grocery stores), and corned beef and pastrami, which are made by soaking beef in a 10-percent salt water brine for several weeks.

Pickling

Pickling was widely used to preserve meats, fruits and vegetables in the past, but today is used almost exclusively to produce "pickles," or pickled cucumbers. Pickling uses the preservative qualities of salt combined with the preservative qualities of acid, such as acetic acid (vinegar). Acid environments inhibit bacteria. To make pickles, cucumbers are soaked in a 10-percent salt water brine for several days, then rinsed and stored in vinegar to preserve them for years.

PASTEURIZING, FERMENTING, CARBONATING

Pasteurizing

Pasteurization is a compromise. If you boil a food you can kill all bacteria and make the food sterile, but you often significantly affect the taste and nutritional value of the food.

When you pasteurize a food (almost always a liquid), what you are doing is heating it to a high enough temperature to kill certain (but not all) bacteria and to disable certain enzymes, and in return you are minimizing the effects on taste as much as you can.

Commonly pasteurized foods include milk, ice cream, fruit juices, beer and non-carbonated beverages. Milk, for example, can be pasteurized by heating to 145 degrees F (62.8 degrees C) for half an hour or 163 degrees F (72.8 degrees C) for 15 seconds.

Ultra high temperature (UHT) pasteurization completely sterilizes the product. It is used to created "boxes of milk" that you see on the shelf at the grocery store. In UHT pasteurization, the temperature of the milk is raised to about 285 degrees F (141 degrees C) for one or two seconds, sterilizing the milk.

FERMENTING

Fermentation uses yeast to produce alcohol. Alcohol is a good preservative because it kills bacteria. When you ferment grape juice you create wine, which will last quite a long time (decades if necessary) without refrigeration. Normal grape juice would mold in days.

CARBONATING

Carbonated water is water in which carbon dioxide gas has been dissolved under pressure. By eliminating oxygen, carbonated water inhibits bacterial growth. Carbonated beverages (soft drinks) therefore contain a natural preservative.

CHEESE-MAKING

Cheese is way of preserving milk for long periods of time. In the process, the milk in cheese becomes something completely unlike milk, but cheese has its own interesting and delicious properties.

Cheese-making is a long and involved process that makes use of bacteria, enzymes and naturally formed acids to solidify milk proteins and fat and preserve them. Once turned into cheese, milk can be stored for months or years.

The main preservatives that give cheese its longevity are salt and acids. The basic steps in cheese making go something like this (for most common cheeses like cheddar):

- First, milk is inoculated with lactic acid bacteria and rennet. The lactic acid bacteria convert the sugar in milk (lactose) to lactic acid. The rennet contains enzymes that modify proteins in milk. Specifically, rennet contains rennin, an enzyme that converts a common protein in milk called caseinogen into casein, which does not dissolve in water. The casein precipitates out as a gel-like substance that we see it as curd. The casein gel also captures most of the fat and calcium from the milk. So the lactic acid and the rennet cause the milk to curdle, separating into curds (the milk solids, fats, proteins, etc.) and whey (mostly water). A gallon of milk (about 8 pounds) yields only about 1.25 pounds of cheese--the weight that is lost is all the water in milk.

- The curds and whey are allowed to soak until the lactic acid bacteria create a lactic acid concentration that is just right. At that point, the whey is drained off and salt is added.
- Now the curds are pressed in a cheese press--lightly at first to allow the escape of the remaining whey, then severely (up to a ton of pressure) to solidify the cheese.
- Finally, the cheese is allowed to age (ripen) for several months in a cool place to improve its taste and consistency. A sharp cheddar cheese has been aged a year or more. During this time, enzymes and bacteria continue to modify proteins, fats and sugars in the cheese. The holes in Swiss cheese occur during ripening--Swiss cheese is ripened in a cool place for several weeks, then put in a warm place (70 degrees F, 21 degrees C or so) for four to six weeks, where special bacteria ferment the remaining lactose and produce carbon dioxide bubbles in the cheese.

As you can see, cheese-making is complicated. It produces a product that preserves milk proteins and sugars with acids and salt.

CHEMICAL PRESERVATION

There are three classes of chemical preservatives commonly used in foods:

- Benzoates (such as sodium benzoate)
- Nitrites (such as sodium nitrite)
- Sulphites (such as sulphur dioxide)

If you look at the ingredient labels of different foods, you will frequently see these different types of chemicals used. Another common preservative that you will commonly see on food labels is sorbic acid. All of these chemicals either inhibit the activity of bacteria or kill the bacteria.

IRRADIATION

Nuclear radiation is able to kill bacteria without significantly changing the food containing the bacteria. So if you seal food in plastic and then radiate it, the food will become sterile and can be stored on a shelf without refrigeration. Unlike canning, however, you do not significantly change the taste or texture of the food when you irradiate it.

The FDA recently approved the irradiation of beef, and the irradiation of chicken has been approved for some time. Irradiation of these meats could prevent many forms of food poisoning. However, many people have a significant problem with the words "nuclear radiation." Therefore, irradiated food is not very common in this country.

3

NUTRITION: FUNDAMENTALS

FOOD CHOICE

Food choice is a complex product of multiple influences including culture, family and the availability of specific foods. Food choices clearly affect the nutritional value of the diet. While foods are primarily purchased by consumers based on taste, cost and convenience, other factors such as nutritional quality, sensory appearance, and individual routines, habits and associations affect food choices as well. In addition, food, agriculture and trade policies, and food production technology, all have tremendous influence on what foods are available for purchase and selection.

MACRONUTRIENTS AND MICRONUTRIENTS

Macronutrients and micronutrients make up the nutritional landscape. Micronutrients (including vitamins, minerals, and phytochemicals) derive their name from the fact that they are needed in relatively small amounts in comparison to the macronutrients-carbohydrates, fats and oils, proteins and amino acids, and water. There are also non-nutrient components of the diet such as soluble and insoluble fibers, which play an important role in human nutrition.

The human body stores macronutrients in a very different proportion from those found in the human diet. While human diets may be 30% fat, 50% carbohydrate, and 20% protein on a daily basis, the body stores 160,000 Calories as fat (about 13.5 kg in a typical 70 kg man), and 54,000 Calories (about 13.5 kg) as protein. Surprisingly, only 1200 Calories of carbohydrate are stored as 300 grams of glycogen in the liver and muscle. These storage fractions are correlated with the portability of the energy stored, with fat being most portable and carbohydrate, requiring water of hydration, being the least portable.

Chemical elements essential to life are classified as major, macrominerals, and microminerals. The macrominerals include calcium, chlorine, magnesium, phosphorus, sodium and sulfur.

The microminerals include arsenic, boron, chromium, cobalt, copper, fluorine, iodine, iron, manganese, molybdenum, nickel, selenium, silicon, tin, vanadium and zinc. These minerals are derived from the diet or dietary supplements. As the diet becomes refined and processed there is the risk of depletion of some of these elements. In more biodiverse diets, the risk of dietary deficiency is reduced.

A mythical 70 kg man contains 45 liters (or kg) of water, and it is the most prevalent of the macronutrients in the body. This 45 liters is divided among intracellular (30 liters), interstitial (12 liters) and blood plasma (3 liters).

A moderately active adult turns over about 2500 ml of water daily, with losses occurring primarily through the urine (50% of daily fluid losses), but also via the skin (25%), lungs (19%) and feces (6%). Body water is derived mainly from beverages (60%), food (30%), and products of metabolism (10%). Excretion is in urine (60%), skin (28%), sweat (8%), and feces (4%).

The metabolism of carbohydrates, fats, and proteins, is interconnected. This interrelationship provides essential flexibility to humans when faced with dietary restriction or starvation. In fact, the body can interconvert to a greater or lesser degree the macronutrients protein, fat and carbohydrate during starvation or overfeeding. Dietary carbohydrates are broken down in the body to glucose, to be used for liver and muscle glycogen stores, and some is taken into the brain and other cells to be broken down to provide energy.

Glucose can also converted to body fat and stored when excess calories are consumed, and, if nitrogen is available, can yield non-essential amino acids. The digestion of fat yields fatty acids and glycerol; some are reassembled as triglycerides which are stored in adipose tissue; others are broken down to provide energy. Dietary protein is broken down into amino acids, which are used to manufacture body proteins, but most amino acids can also be converted to glucose when energy is needed, and if taken to excess will be stored as body fat. When there is a surplus of amino acids, or if there is inadequate carbohydrate or fat to meet energy needs, amino acids can also provide energy. Of the energy-containing nutrients, fat provides the most energy by weight.

PROTEIN

Essential Amino Acids	Non-essential Amino Acids
Histidine	Alanine
Isoleucine	Arginine
Leucine	Asparagine
Methionine	Aspartic Acid
Lysine	Cysteine
Phenylalanine	Glutamic Acid
Threonine	Glutamine
Tryptophan	Glycine
Valine	Proline
	Serine
	Tyrosine

Proteins are composed of amino acids. These are classified as essential or non-essential. Essential amino acids must be obtained from the diet. Non-essential amino acids can be synthesized in the body from essential amino acids. The quality of dietary proteins is based on their content of essential and non-essential amino acids. The proteins with highest utilization efficiency are said to have the highest biological value.

The protein-digestibility-corrected amino acid score (PDCAAS) method of evaluating the quality of a protein compares the amino acid contents of a protein with human amino acid requirements and corrects for digestibility. Amino acid scoring is a method of evaluating protein quality by comparing a test protein's amino acid patterns with that of a reference protein. Once the amino acid score is derived, it is compared against the amino acid requirements of preschool-aged children. The rationale behind using the requirements of this age group is that if a protein will effectively support a young child's growth and development, it will meet or exceed the requirements of older children and adults.

Protein Digestibility Corrected Amino Acid Score Values

Casein (milk protein)	1.00
Egg White	1.00
Soybean isolate	0.99
Beef	0.92
Pea flour	0.69
Kidney beans	0.68
Garbanzo beans	0.66
Pinto beans	0.63
Rolled oats	0.57
Lentils	0.52
Peanut meal	0.52
Whole wheat	0.40

LIPIDS

Lipid classes include fatty acids, fatty acid derivatives, and sterols. Fatty acids act primarily as metabolic fuel. The fatty acid derivatives are triglycerides, phospholipids and cholesteryl esters. Triglycerides are the transport form of lipids in the body, and also the form in which fatty acids are stored. Phospholipids the primary component of cell membranes, and cholesteryl esters are the storage form of cholesterol. The sterol class of lipids includes cholesterol, bile acids and hormones. Cholesterol is a component of cell membranes, bile acids function in fat digestion, and hormones function in metabolic regulation.

Fatty acids are an essential component of the diet, albeit at a low level of 10% of the total calories in the diet to assure that

6% comes from a combination of linoleic and linolenic acids, essential fatty acids. Fats impart many qualities to foods, including flavor, texture, aroma and appearance. Many of the compounds that provide flavor are fat soluble, and fat transports these substances to sensory cells in the mouth, providing taste and aroma.

Fat also provides a creamy sensation in foods such as ice cream, and tenderness in meats which are marbled with fat. In baked goods, fats provide structure and tenderness. In response to consumer demand, food manufacturers have introduced thousands of fat-free and low fat foods to the marketplace, which has created the challenge of reducing food fat content while at the same time preserving the taste and textural qualities that fat provides.

CARBOHYDRATES

Carbohydrates are classified as simple or complex. Simple carbohydrates include the monosaccharides (glucose, fructose and galactose) and disaccharides (maltose, lactose and sucrose). Glucose is the primary carbohydrate in the bloodstream; fructose is the primary sugar in fruits. Galactose rarely appears free in nature, but in combination with glucose to form the dissaccharide, lactose, which is the predominant sugar in milk.

Sucrose (glucose + fructose) is table sugar, and maltose (glucose + glucose) is the primary disaccharide in beer. Complex carbohydrates are composed of many monosaccharides linked together to form starches and fibers. An anomalous group classified as complex by labeling law are the polysaccharides. These dissolve in the stomach and essentially act like simple carbohydrates except for their effects on intestinal osmolality which may influence bowel motility.

Fiber is an important non-nutrient constituent of foods such as cereals, grains, fruits and vegetables. Fibers can be classified as soluble or insoluble. Soluble fiber is present in apples, citrus, oats, barley and legumes, and delays gastric emptying time, delays glucose absorption, and stimulates colonic fermentation. Insoluble fiber, found in whole grain breads and cereals, wheat bran, and vegetables, accelerates GI transit time, delays glucose absorption, and increases fecal weight.

Fiber has a number of actions in the colon. In addition to the binding of certain drugs, fiber can have effects on drug metabolism through the actions of colonic bacteria which can in turn be influenced by the fiber content of the diet. Studies have demonstrated that dietary differences correlate with differences in colonic bacteria species. Some of these species differences have functional implications for nutrient and hormone metabolism. Finally, fiber can affect transit time of nutrients in the colon which in turn can affect absorption of drugs through the colonic mucosa.

VITAMINS AND MINERALS

Vitamins are organic compounds which largely serve as coenzyme factors. Their name derives from "vita" from the Latin for life and "amine" signifying a protein, since the early vitamin researchers assumed these factors must be proteins. In fact, the vitamins are defined operationally by their necessity for the maintenance of life. Some vitamins such as vitamins A and D are really steroid hormones unrelated to the many enzyme co-factors found among the vitamins.

The fat soluble vitamins are A, D, E, and K. Because they are stored by the body, fat soluble vitamins are less readily excreted than water soluble vitamins, and are needed in periodic doses. The water soluble vitamins include the B vitamins [B1(thiamin), B2 (riboflavin), B3 (niacin) B6(pyridoxine), folic acid and B12 (cyanocobalamin)] and vitamin C (ascorbic acid).

These vitamins are absorbed directly into the blood stream and freely circulate. Since the kidneys detect excesses in the blood and excrete them into the urine, water soluble vitamins are needed in more frequent doses.

In addition to these vitamins, fruits and vegetables provide a large number of phytochemicals, including carotenoids, isoflavones, terpenoids, isothiocyanate, organic sulfides, and polyphenols which have various physiological effects. The functions of many of these compounds have not been fully elucidated.

Vitamin D is different from all the other nutrients in that it can be synthesized from 7 dehydrocholesterol in the skin in the presence of sunlight. Ergosterol can also be ingested from plants and converted to active forms of vitamin D. The vitamin functions as a hormone, with the intestines, kidneys and bones as target organs.

These organs respond to vitamin D by making calcium available for bone growth. Ultraviolet rays from the sun convert 7-dehydrocholesterol into previtamin D3. The liver hydroxylates the compound to 25-hydroxy vitamin D3, and within the kidney further hydroxylation yields either active (1,25 dihydroxy vitamin D3) or inactive (24, 25 dihydroxyvitamin D3) forms of the vitamin. This important branch point regulation is critical in regulation of calcium metabolism.

Vitamin B 12 illustrates an example of complex regulation of vitamin absorption. After ingestion, the vitamin B12 present in protein foods is released from the proteins by hydrochloric acid and pepsin in the stomach. Absorption of the vitamin requires another protein (intrinsic factor), secreted by parietal cells in the stomach, to bind to B12 and travel to the ileum where the complex is absorbed.

Many vitamin B 12 deficiencies reflect inadequate absorption, rather than poor intake. Inadequate absorption can occur due to lack of hydrochloric acid, or lack of intrinsic factor. Many older adults develop atrophic gastritis, in which inadequate hydrochloric acid is produced, and vitamin B12 supplements are often recommended to this population for this reason.

Vitamin B12 and folate are interdependent for activation. Since vitamin B12 is required to convert folate to its active form, the most obvious vitamin B12 deficiency symptom is anemia of folate deficiency.

Minerals are inorganic compounds and are classified as either major minerals or trace minerals, depending upon the amounts present in the body. The major minerals (calcium, phosphorus, potassium, sulfur, sodium, chloride and magnesium) are present in amounts larger than 5 grams. The major minerals provide structure to the body and help to maintain fluid balance.

The trace minerals function primarily as coenzymes in energy metabolism. The trace minerals (iron, zinc, copper, manganese, iodine and selenium) are found in the human body in amounts less than 5 grams. Some foods contain binders, such as phytic acid and oxalic acid, which can form complexes with minerals and prevent their absorption. Iron metabolism illustrates a carefully regulated system at the intracellular level. When there is deficient iron, the cell triggers production of transferrin receptors at the cell membrane to transport iron into the cell. When there is excess iron available, the cell triggers the production of ferritin, an iron-storage protein.

The iron response element (IRE) is responsible for both of these environmental signals at the molecular genetic level. Most iron is recycled, but some is lost via sweat, skin and urine.

PHYTOCHEMICALS

Phytochemicals are nonnutrient compounds in plants that have biological activity in the body. Some examples of these substances are carotenoids in deeply pigmented fruits and vegetables, allyl sulfides in garlic and onions, sulforaphane in cabbage family plants, and flavonoids in fruits, vegetables, and tea.

In addition to imparting flavor and color to foods, phytochemicals can have profound physiological effects, including the suppression of the development of cancer. There are several modes of action of phytochemicals. Some, such as carotenoids and flavonoids, act as antioxidants. Others, such as limonene in citrus fruits and allyl sulfides in garlic and onions, trigger enzyme production to facilitate carcinogen excretion. Caffeic acid in fruits triggers enzyme production to make carcinogens water-soluble, facilitating excretion.

There are many active hormone-like substances found in plants such as isoflavones in soy protein. The isoflavones, genistein and daidzein, are selective estrogen response modifiers. They are proestrogenic in bone and brain, and anti-estrogenic in breast and uterus. Isoflavones are also antioxidants, and tyrosine kinase inhibitors capable of inhibiting breast and prostate tumor cell growth in vitro.

XENOBIOTICS

Xenobiotics are external elements providing a signal to the cell. Often these xenobiotics have a precursor form which requires activation by oxidation. These reactions are carried out by so-called drug metabolizing enzymes (DME's).

Other DME's carry out inactivation of the oxygenated ligand providing a regulatory mechanism. Apoptosis, or cell death, differentiation, and neuroendocrine functions can be carried out in response to the expression of specific genes triggered by active oxygenated ligands derived from xenobiotics.

Potentially carcinogenic dietary constituents and potentially protective dietary constituents interact in pathways of DNA oxidation and damage leading to carcinogenesis. These pathways are well-studied for smoking related carcinogens and there is emerging evidence on the applicability of this common pathway to drugs and food constituents.

Potentially carcinogenic dietary constituents and potentially protective dietary constituents also co-exist in the food supply. While pesticides are considered a potential source of carcinogens, the protective substances found in foods may counteract the harmful effects of other toxins.

GENE-NUTRIENT INTERACTION

While man evolved on earth about 50,000 to 100,000 years ago, the drug-metabolizing enzymes were developed hundreds of millions of years ago in simple bacteria. For example, cytochrome P450 has over 150 isoforms in different species. Agriculture only evolved 10,000 years ago and modern western diets have been here only a few hundred years. There is a gene-environment imbalance for individuals eating a typical high fat diet.

The Pima Indians of Northern Mexico and Southern Arizona demonstrate the impact of nutrition on genetic expression of chronic diseases such as obesity and non-insulin dependent diabetes mellitus (NIDDM). Pimas in Arizona have the highest reported prevalence of obesity and non-insulin dependent diabetes, while the incidence is low in Pimas living in Northern Mexico.

Pima Indians living in the mountains of Northern Mexico were separated some 700 to 1000 years ago from the Pimas living today in Arizona. The Northern Mexico inhabitants continue to live a traditional lifestyle, eating a diet with much less animal fat and more complex carbohydrates, and have higher activity levels, than Pimas in Arizona. This population demonstrates increased body weight, height (which is a biomarker of prepubertal nutrition), increased body mass index, increased cholesterol and increased incidence of NIDDM.

The "thrifty gene" hypothesis states that genes for conserving energy, while adaptive in a traditional environment, are responsible for the high prevalence of obesity and diabetes in individuals exposed to overnutrition and underactivity

FOOD PYRAMIDS AND DIETARY RECOMMENDATIONS

Recommendations for dietary intake among populations often take the form of pyramids, with the base of the pyramid determining the basis of the diet. In the USDA pyramid developed in the late 1980's the base consists of cereals and grains without an emphasis on the fiber content of those grains. The fruits and vegetables appear on the second level and are two separate groups. In a new pyramid developed in 1997 by the UCLA Center for Human Nutrition, a modified plant-based diet is recommended.

The base consists of 5 to 11 servings of fruits and vegetables to provide unique phytonutrients for chronic disease prevention. The second level is 6 to 9 servings of high fiber cereals and grains to provide the benefits of fiber from this level as well as from fruits and vegetables. The protein level is made up of low fat protein choices including protein from the plant kingdom (beans and rice or soybean protein) and the animal kingdom (egg whites, breast of chicken and turkey, low-fat fish and seafood, and non-fat milk products). The top tier of the USDA pyramid has only dots representing fats and sweets with the mixed message "use sparingly".

The top tier of the new California Pyramid emphasizes taste enhancers including olives, avocados, garlic, onion, nuts, cheese, chili peppers, and as much monounsaturate or omega-3-rich oils as needed to enhance taste. The overall fat recommendation of the USDA pyramid is 30% or less of calories from fat, while the new pyramid reflects the decreased fat intake in the population by recommending 20% or less of calories from fat. This exercise illustrates how the science of nutrition can be used to influence dietary recommendations

BIOCHEMISTRY OF OXIDANT STRESS IN HEALTH AND DISEASE ANTIOXIDANTS

The human genome dates back over 40,000 years ago when man was a "hunter-gatherer" and consumed a diet consisting of over 90 percent plant-derived foods compared to about 30 percent in the modern post-industrial era. Our modern diet is largely derived from animal sources which are higher in fat, lower in fiber, and poorer in micronutrients especially antioxidants.

While our modern diet has virtually eliminated malnutrition in the majority of the population, there is an epidemic of obesity (affecting 35% of Americans in the latest Department of Agriculture survey in 1989) and an increased incidence of heart disease, breast cancer, prostate cancer, and colon cancer by comparison with populations eating less meat and more fruits, vegetables, cereals and grains. The increasing incidence of chronic diseases and obesity in Japan which has adopted a more Western diet over the last 30 years is strong evidence of the importance of nutrition in disease prevention.

There are three levels at which the diet of an individual can be assessed: (1) the overall caloric content, and macronutrient profile (i.e. protein, carbohydrate, and fat composition); (2) the vitamin and micronutrient adequacy for prevention of deficiency diseases; and (3) the adequacy of antioxidants including carotenoids, vitamin C, vitamin E, and selenium for prevention of heart diseases and common forms of cancer.

OXIDANTS AND ANTIOXIDANTS

The process of oxidation occurs in any oxygen-rich environment where substrates are exposed to ultraviolet light or heat. Examples of commonly occurring chemical oxidation processes include an apple turning brown following brief exposure to air, a car rusting in the open air, and the formation of rancid fats in poorly preserved foods.

These processes are all mediated by highly reactive oxygen radicals (oxygen atoms with a single unpaired electron). In order to understand the mechanisms by which antioxidants can prevent chronic diseases, it is necessary to understand the actions of free radicals and how antioxidants limit free radical reactions.

Oxygen involved in the oxidation of substrates to produce energy in normal metabolic processes can produce oxygen radicals. They can have a beneficial roles as in phagocytes where they protect against bacteria and parasites.

However, if natural antioxidant mechanisms are not adequate to quench excess oxygen radicals then they can react with cell structures. Metabolism is not the only source of free radicals. Environmental pollutants are sources for free radicals including nitrogen dioxide, ozone, cigarette smoke, radiation, halogenated hydrocarbons, heavy metals and certain pesticides. Alcohol consumption can induce oxidative reactions in the liver.

Certain chemotherapeutic agents including doxorubicin, cyclophosphamide, 5-fluorouracil, methotrexate, and vincristine can produce oxygen radicals at doses used in cancer patients. Increased physical activity can generate free radicals as the result of increased oxygen consumption during exercise. Oxygen radicals in the human body react with proteins, lipids, carbohydrates and nucleotides.

When free radicals attack polyunsaturated fats in the presence of oxygen, lipid peroxides are formed in the a chain reaction that results in amplification of the original oxidative damage. For example, low density lipoprotein that is oxidized is more easily taken up by macrophages in the endothelial wall of blood vessels promoting atherogenesis. A DNA base oxidation product (8-hydroxydeoxyguanosine) can be detected in the urine of humans exposed to oxidant stresses raising the possibility that oxidation can alter genetic information. These oxidative processes have been associated with degenerative changes occurring with aging, and the development of cardiovascular disease and cancer. Free radical reactions occur continuously in living cells, but most of the changes resulting are repaired. Damage that escapes these repair systems may accumulate over long periods of time and play a role in degenerative diseases.

The body's susceptibility to oxidant damage is thought to depend on the balance between the extent of pro-oxidant stress and the antioxidant levels of body tissues. Most antioxidants have a large number of alternating double bonds which can act as electron traps. In some cases, this quenching reaction can lead to increased oxidation.

This occurs when a polyunsaturated fat neutralizes an oxygen radical but becomes a fatty acid radical which then attacks another lipid leading to a chain reaction. Other antioxidants can also act as pro-oxidants after quenching an oxygen radical. On balance, a number of studies have shown that various antioxidants act as a cooperative system of antioxidant defense. Vitamin C quenches free radicals in aqueous systems, but also regenerates cellular vitamin E which helps to control lipid peroxidation.

Beta-carotene also traps free radicals in concert with vitamin E. The selenium-containing enzyme glutathione peroxidase destroys peroxides before they can damage cell membranes and interacts synergistically with vitamin E. In a number of animal studies, the administration of antioxidants ameliorated damage from experimental oxidant stress. Furthermore, the antioxidant requirement in these studies was directly proportional to the increased tissue concentrations of free radicals. While not proven, promotion of antioxidation through consumption of antioxidant-rich fruits and vegetables as well as dietary supplements is a habit many Americans consider health-enhancing.

On the basis of population studies and animal studies, a number of governmental agencies including the National Research Council, the National Cancer Institute and the United States Department of Agriculture have recommended that Americans eat at least five servings a day of fruits and vegetables in part to increase the intake of beneficial antioxidants. Recent surveys indicate that only a small fraction of the general population follows this advice, leading some nutrition authorities to recommend dietary supplementation or food fortification with antioxidants to reduce chronic disease incidence. While there remains some uncertainty about the long term effects of antioxidant supplementation, there is accumulating evidence that the practice of dietary supplementation may be beneficial.

Carotenoids are found in green, yellow, and orange vegetables and some fruits. The commonly occurring carotenoids include beta and alpha carotene, lycopene, cryptoxanthin, lutein, and zeaxanthin. These all act as antioxidants, but only beta carotene can be converted to vitamin A, and only beta carotene is available as a dietary supplement. Beta carotene is classified as a generally safe natural food coloring.

The only toxic effect in man is yellowing of the skin at doses of greater than 20 mg per day. It has been estimated that individuals eating five servings of fruits and vegetables per day would obtain about 6 mg of beta carotene from their diets. The average intake in Americans is about 3.5 mg per day currently. There is clearly a large gap between the recommended intakes and the level of toxicity making supplementation with up to 10 mg generally safe.

Vitamin E is a generic term that includes entities exhibiting the biological activities of d-alpha tocopherol. In nature, eight substances have vitamin E activity: d-alpha, d-beta, d-gamma, and d-delta tocopherols and d-alpha, d-beta, d-gamma, and d-delta tocotrienols. D-alpha tocopherol has the highest biologic activity in the vitamin E assay but all of these are antioxidants. There is also a synthetic form of vitamin E: d,l-alpha tocopherol.

It bioactivity in the vitamin E assay is reduced (1.0 vs. 1.49). Vitamin E is present in small amounts in a large number of foods including vegetable oils (soybean, sunflower, and corn oil), wheat germ, whole grains, egg yolk, nuts, sunflower seeds, green vegetables, milk fat, and liver. The Seventh RDA Committee in 1968 set the RDA for vitamin E at 20 mg (30 IU), but found that this was difficult to attain through normal diets without supplementation. Subsequent editions of the RDA halved the requirement to 10 mg (15 IU).

It is estimated that 64 percent of the vitamin E intake of the American diet is supplied by salad oils, margarine, and shortening. Eleven percent is supplied by fruits and vegetables, and about 7 percent by grains and grain products. The average amount supplied by the American diet is estimated to be 7 to 9 mg (10.4 to 13.4 IU). Vitamin E in amounts considered to be protective are not easily consumed through the diet. For example, a quart of corn oil contains 200 IU of vitamin E but over 7,700 Calories.

Some commonly consumed dietary supplements with antioxidant activity are referred to as antioxidant vitamins although not all components of these combinations are vitamins. These preparations most commonly include a combination of vitamin E, vitamin C, and beta carotene. Beta carotene is an antioxidant that is converted to vitamin A only under conditions of vitamin A deficiency. Vitamin E and vitamin C only have antioxidant properties at doses well above the doses required to prevent vitamin deficiencies. Other carotenoids such as lycopene (the red pigment found in tomatoes) are not converted to vitamins but act as antioxidants. Certain antioxidants such as ubiquinone (also known as Coenzyme-Q10) and glutathione are synthesized in the body and are not essential dietary constituents. Most of the studies cited below have considered the three most commonly consumed antioxidants-Vitamin E, Vitamin C, and beta carotene.

Other antioxidants such as the flavinoids may have other pharmacologic actions including anti-hormonal and anti-growth factor effects at a cellular level. For example genistein, an isoflavone found in soybean protein, binds to the type II estrogen receptor and interferes with EGF receptor tyrosine kinase phosphorylation in breast cancer cells. While there is much that is not known about antioxidants, there is evidence accumulating on the preventive role of antioxidants in the two most prevalent chronic diseases-heart disease and cancer.

ANTIOXIDANTS IN HEART DISEASE

Approximately 70 million Americans have one or more forms of cardiovascular disease including hypertension, coronary heart disease, stroke, and rheumatic heart disease. Every year it is estimated that 1.5 million Americans will have a heart attack, with approximately 500,000 to 600,000 deaths resulting. The estimated cost of coronary heart disease in 1993 was 51.6 billion dollars. Data from numerous studies indicate a beneficial effect of supplemental vitamins C, E, and beta carotene in significantly reducing coronary events, reperfusion injury, platelet aggregation, and low-density lipoprotein (LDL) oxidation. Two large-scale epidemiologic studies of men and women show vitamin E consumption at levels greater than 100 IU per day to be associated with a reduced incidence of coronary artery disease. In the Nurses' Health Study, women who consumed more than 3 times the RDA of vitamin E had a 34% lower risk of heart attacks than women who consumed lesser amounts. Similar results were found in the Health Professionals Study which involved men.

These epidemiologic studies, combined with evidence in laboratory studies that concentrations of vitamin E and beta carotene that can be achieved with vitamin supplementation resulted in decreased susceptibility to oxidation of LDL isolated from plasma following in vitro challenge with pro-oxidants (copper or iron), have led many cardiologists to recommend that their patients take antioxidant supplements. While additional studies are still required to reach a significant scientific agreement on the benefits of antioxidants according to the Food and Drug Administration, many physicians are advising their patients to take supplements. The determination of compliance with this advice as well as confirmation of the achievement of steady-state levels in the range observed to be beneficial in epidemiologic studies will require measurement of blood levels of antioxidants.

ANTIOXIDANTS IN CANCER

The common forms of cancer, including breast, colon, and prostate cancer, are the result of genetic-environmental interactions. While a small minority (5-10%) of patients have inherited forms of cancer due to alterations in the genes of the germ cell line, cancers are all assumed to have genetic changes at the somatic cell level. These genetic changes lead to unregulated growth through activation of growth-promoting genes (oncogenes) or inactivation of tumor suppressor genes.

In some cases oncogenes code for growth factor receptor proteins and growth factors. Reactive oxygen radicals are thought to damage biologic structures and molecules including lipids, protein, and DNA, and there is evidence that antioxidants can prevent this damage.

For example, vitamin C has three proposed mechanisms of action in chemoprevention: 1) incorporation of ascorbic acid into the hyaluronidase inhibitor system; 2) prevention of the formation of nitrosamines in the gastrointestinal tract; and 3) inhibiting the potency of chemical carcinogens by enhancing the activity of the detoxifying cytochrome P-450 system.

Beta carotene and retinol have been shown in vitro and in animal studies to suppress carcinogenesis. Vitamin E in cell culture has been shown to reduce expression of certain oncogenes (H-ras and c-myc) in tumor cells, and to inhibit cell proliferation in other cell lines. Selenium has been shown to prevent the metabolism of chemical carcinogens necessary for their activation. Rats fed a diet high in beta carotene, vitamin C, or selenium were demonstrated to have a reduced incidence of carcinogen-induced pancreatic cancer.

Although the antioxidant properties of vitamin C, vitamin E, and beta carotene have been considered the primary mode of action, other mechanisms may also be involved in their chemopreventive actions. These include effects on the immune system, on modulation of growth factors and growth factor receptors, and on intercellular interactions. Epidemiologic studies provide evidence on the relationship of the intake of antioxidants to the incidence of lung cancer. In Finnish studies carried out over periods of five to twenty years, there was an inverse relationship between serum levels of vitamin E and other antioxidants and the risk of cancer at several anatomic sites. In a study of 15,093 women ages 15-99 years and initially free of cancer, cancer was diagnosed in 313 women over an eight year follow-up. Low serum levels of vitamin E strongly predicted the risk of epithelial cancers, but demonstrated only slightly elevated risks of cancers in reproductive organs exposed to estrogens. Therefore, while the effects of antioxidants can be demonstrated in animal studies, human studies are more complex due to the known effects of obesity increasing estrogen levels in women, and other factors in the Western diet such as reduced fiber intake.

Smoking sometimes also appears to override the protective effects of antioxidant vitamins in some studies. A randomized double-blind prospective controlled trial carried out for five to eight years in 29,133 male smokers studied the effects of 50 mg synthetic vitamin E alone or in combination with 20 mg synthetic beta carotene. Among the 876 cases of lung cancer observed there was no decrease observed in the men supplemented with vitamin E. There were 52 fewer cases of prostate cancer. Unexpectedly, the incidence of lung cancer was 18% higher and the mortality due to lung cancer was higher in the beta carotene supplemented group.

This result was unexpected and runs counter to the body of experimental evidence and a recent supplementation trial in China in which 29,584 adults ages 40 to 69 years were provided a variety of vitamin supplements including vitamin A, riboflavin, vitamin C, and a combination of beta carotene, selenium, and vitamin E. The combination of selenium (50 mcg), vitamin E (30 mg) and beta carotene (15 mg) was associated with a 13% decrease in cancer mortality. The major effect was on gastric cancer (21% reduction) which is the most common cancer in this population.

Clearly, much further research needs to be done on the effects of antioxidants, but the above studies indicate a potential protective effect on cancers of the aerodigestive system. In the presence of other major etiologic factors (e.g. estrogen for breast cancer, or smoking for lung cancer), the beneficial effects of antioxidants are not demonstrable. In patients at increased risk of cancer, including the population that has undergone primary treatment of cancer, antioxidant supplementation is being recommended by many physicians.

FUTURE DIRECTIONS OF DEVELOPMENT

Antioxidants work by supplementing significant host defenses against oxidant stress. The measurement of the host response to oxidant stress remains problematic. There are a number of proposed markers of oxidant stress including measurement of lipid oxidation products such as conjugated dienes, malondialdehyde or thiobarbituric acid reactive substances (TBARS) in blood or urine; modified DNA bases (8-hydroxydeoxyguanosine) and/or DNA adducts in peripheral blood cells or urine; vitamin E or vitamin C levels in blood fractions; catalase or dismutase levels in blood fractions; volatile gases such as pentane or ethane in expired breath; total peroxyl radical trapping antioxidant power of serum (TRAP assay); auto-oxidative (non-cyclooxygenase-derived) eicosanoids in plasma; and the in vitro oxidation of blood fractions such as LDL.

Future refinement of these methods with clinical correlation will provide important information for designing supplement strategies for disease prevention.

While a number of government agencies recommend that Americans consume more than 5 servings per day of fruits and vegetables it is clear that only a minority are able to follow this advice. Supplementation of usually consumed foods with antioxidants is another potential strategy.

It is already true that breakfast cereal is a major dietary source of folic acid for Americans who consume too little of green leafy vegetables to obtain the Recommended Dietary Allowances. It may be that foods can be developed which provide protective amounts of antioxidants in lieu of the capsules currently available. Additional research on the differences among the various antioxidants and how they work must also be done and reduced to dietary changes for the population.

Finally, normal ranges for different age groups referenced to dietary intake need to be developed. It is clear that antioxidant levels in the blood are influenced by dietary intake including seasonal variation in the intake of fruits and vegetables. A workshop sponsored by the Food and Drug Administration in 1993 outlined a pro-active strategy to be utilized in evaluating data on the positive impact of nutrient and non-nutrient antioxidants on human health. The World Cancer Research Fund and the American Institute of Cancer Research issued a 740 page analysis of international data supporting new dietary guidelines which indicate that between 400 and 800 gm per day of vegetables may be preventative. The overall strategy of these groups includes engaging the scientific community through advisory councils workshops and other mechanisms. It is likely that much new information on the benefit of antioxidants will be developed in the next few years providing clinicians with improved guidelines for following the antioxidant status of their patients at risk of cardiovascular diseases and cancer.

4

BASICS OF NUTRIENT METABOLISM

PROTEIN STORES AND THEIR ROLE IN SURVIVAL

In the average 70 kg man, the largest store of calories is in the form of fat in adipose tissue with approximately 135,000 Calories stored 13.5 kg of adipose tissue. This storage compartment can be greatly expanded with long-term overnutrition in obese individuals. There are approximately 54,000 Calories stored as protein both in muscle and viscera. Only half of these calories can be mobilized for energy, since depletion below 50 percent of total protein stores is incompatible with life.

In addition to being an energy source, protein plays a functional role in many organs including the liver, and depletion is associated with impaired immunity to infection. In fact, the most common cause of death in an epidemic of starvation is typically simple bacterial pneumonia. Conservation of protein is an adaptation tightly linked to survival during acute starvation.

There are only 1200 Calories stored as carbohydrate in liver and muscle glycogen. There are clear adaptive advantages to storing calories as fat, since fat can provide more energy per gram than carbohydrate or protein. However, since carbohydrate stores are so small, they are depleted in three days of uncomplicated starvation or sooner under conditions of increased energy expenditure. This dependence on fat and protein stores in starvation requires metabolic adaptations to minimize the loss of protein stores, and a shift to metabolic pathways predominantly utilizing the large fat stores available.

THE ADAPTATION TO STARVATION

The postabsorptive period is defined as 8 to 16 hours after eating, and has been operationally defined as the timepoint after an overnight fast when a number of hormonal determinations can be made under standard conditions. It can be thought of as a period of very early adaptation to starvation. During this period, the primary metabolic priority is the provision of adequate glucose for essential functions of the brain, red blood cells, peripheral nerves, and renal medulla. During this postabsorptive phase, insulin levels fall as blood glucose falls from a range of 4 to 5 mmol/liter to 3 to 4 mmol/liter. Glucose is released from the liver into the circulation via glycogenolysis of stores accumulated after feeding under the influence of insulin.

The fall in glucose levels is associated with the depletion of glycogen stores. Skeletal muscle does not release glucose from stored glycogen directly into the circulation, because myocytes lack the required enzyme, glucose-6-phosphatase. However, muscle releases lactate and amino acids such as alanine, which can enter the circulation and are converted to glucose in the liver via gluconeogenesis. Glucagon in the presence of lowered insulin concentrations promotes gluconeogenesis during the postabsorptive period. In addition, glucagon in the presence of lowered insulin levels promotes lipolysis. As the stored triglyceride in adipocytes is mobilized as free fatty acids, those tissues that do not require glucose as their primary fuel (e.g. skeletal muscle) begin to oxidize free fatty acids. These changes during the early postabsorptive period act to increase free fatty acid oxidation in order to spare protein breakdown. During the first few days of starvation, free fatty acid concentrations increase from a range of 0.5 to 0.8 mmol/liter up to 1.2 to 1.6 mmol/liter, and plateau thereafter as starvation is prolonged.

These free fatty acids circulate bound to albumin, and are oxidized in the liver to water soluble ketone bodies, including acetoacetate and beta-hydroxybutyrate. Following 4 to 6 weeks of uncomplicated starvation in obese subjects with more than adequate triglyceride stores, acetoacetate concentrations rise 25-fold and beta-hydroxybutyrate concentrations rise 100-fold from the levels observed in the postabsorptive phase. These are the largest fluctuations seen in any circulating fuel with prolonged starvation.

Protein synthesis and catabolism has been estimated to account for approximately 40 percent of resting energy expenditure. In addition, the changes in protein metabolism are critical to maintaining the body cell mass during starvation which directly impacts survival. Plasma amino acids measured in venous blood give non-specific indications of the adaptations taking place in protein metabolism during the course of starvation. In addition, the excretion of protein from the body as urinary urea nitrogen expressed as nitrogen balance provides further insights into overall protein nutriture.

The basic mechanisms underlying these adaptive changes in protein synthesis and degradation are still not completely understood. Proteolysis occurs in cellular lysosomes via autophagy. This process is stimulated by a shortage of critical regulatory amino acids including phenylalanine, tryptophan, methionine, leucine, tyrosine, glutamic acid, proline and histidine. While not conclusively established, it appears that decreased concentrations of specific amino acyl transfer RNA's for these amino acids trigger proteolysis. In terms of protein synthesis, there is a decrease in the amount and activity of RNA subunits involved in initiation, elongation, and termination of protein synthesis. Insulin is the primary hormone known to regulate protein metabolism. Insulin deficiency leads to net protein breakdown, and hyperinsulinemia under euglycemic conditions inhibits proteolysis.

There is also evidence that glucagon participates in this regulatory process by stimulating splanchnic proteolysis. Plasma cortisol levels are increased for several hours and inhibit protein synthesis while increasing protein breakdown. Elevations in epinephrine, previously thought to increase protein breakdown lead to decreases in the rate of whole body protein breakdown. Growth hormone has been shown to increase protein synthesis but to oppose insulin's antiproteolytic effects. The role of insulin-like growth factor I is still not established. Recent studies have demonstrated that plasma amino acid levels and amino acid availability play an important role in modulating the rate of protein breakdown. The magnitude of these amino acid-mediated antiproteolytic effects were equivalent to those of insulin.

IMPACT OF PROTEIN CONSERVATION

The impact of the adaptation to a fat fuel economy is reflected in the rapid changes in urinary nitrogen excretion reflecting net protein sparing through two processes. First, there is less protein breakdown. It has been estimated that protein synthesis decreases in the whole body by 40 per cent between the postprandial and postabsorptive phases with a further decrease over the first several days of starvation. Secondly, there is increased reutilization of nitrogen evidenced by decreased urea formation in the liver through the arginine-citrulline cycle. In obese subjects fasting for 7 days, protein breakdown and urinary urea nitrogen excretion decrease in parallel. Overall nitrogen is conserved so that nitrogen excretion decreases from 12 g/day in the postabsorptive state to 5 g/day seven to ten days later. This decrease translates into a decrease in muscle protein breakdown from 75 g/day to 20 g/day. Based on theoretical calculations of the time necessary to reach the critical 50 per cent of body cell mass, survival is extended through these adaptations from approximately 60 days to over 260 days provided that adequate fluid and electrolytes are administered.

Within seven to ten days of starvation, there is a marked adaptive decrease in energy expenditure. Normally, resting energy expenditure is proportional to lean body mass. However, after seven to ten days of starvation, there is a twenty per cent decrease in resting energy expenditure, at a time when lean body mass has decreased by less than five per cent. Changes in the peripheral metabolism of thyroid hormones occur which may contribute significantly to the observed decrease in energy expenditure.

Among these changes, there is less production of triiodothyronine, the most metabolically active thyroid hormone, via a decreased activity of 5'monodeiodinase in the liver and other peripheral tissues. There is a reciprocal rise in reverse triiodothyronine, an inactive metabolite, while thyroxine levels remain constant. The overall decrease in energy expenditure with starvation is an adaptive change which results in a decreased rate of whole body lipolysis, proteolysis, and gluconegenesis. Aerobic exercise in obese dieters does not reverse this adaptive change in energy regulation.

There is a good correlation between the adaptive hormonal changes which occur during starvation, and the decrease in whole body protein breakdown which occurs as a result.

REAL MEANING OF MICROBIOLOGY

The focus of Food Microbiology is on the detection and analysis of foodborne spoilage microorganisms. Food microbiology is the study of food micro-organisms; how we can identify and culture them, how they live, how some infect and cause disease and how we can make use of their activities.

Microbes are single-cell organisms so tiny that millions can fit into the eye of a needle. They are the oldest form of life on earth. Microbe fossils date back more than 3.5 billion years to a time when the Earth was covered with oceans that regularly reached the boiling point, hundreds of millions of years before dinosaurs roamed the earth.

The field of food microbiology is a very broad one, encompassing the study of microorganisms which have both beneficial and deleterious effects on the quality and safety of raw and processed foods. Food science is a discipline concerned with all aspects of food-beginning after harvesting, and ending with consumption by the consumer. It is considered one of the agricultural sciences, and it is a field which is entirely distinct from the field of nutrition. In the U.S., food science is typically studied at land-grant universities.

Examples of the activities of food scientists include the development of new food products, design of processes to produce these foods, choice of packaging materials, shelf-life studies, sensory evaluation of the product with potential consumers, microbiological and chemical testing. Food scientists in universities may study more fundamental phenomena that are directly linked to the production of a particular food product. Food scientists are generally not directly involved with the creation of genetically modified (bio-engineered) foods. Some of the subdisciplines of food science: Food safety, Food engineering, Product development, Sensory analysis, Food chemistry.

The primary tool of microbiologists is the ability to identify and quantitate food-borne microorganisms; however, the inherent inaccuracies in enumeration processess, and the natural variation found in all bacterial populations complicate the microbiologists job.

Without microbes, we couldn't eat or breathe.

Without us, they'd probably be just fine.

Understanding microbes is vital to understanding the past and the future of ourselves and our planet.

Archaea look and act a lot like bacteria. So much so that until the late 1970s, scientists assumed they were a kind of "weird" bacteria.

Then microbiologist Carl Woese devised an ingenious method of comparing genetic information showing that they could not rightly be called bacteria at all. Their genetic recipe is too different. So different Woese decided they deserved their own special branch on the great family tree of life, a branch he dubbed the Archaea.

Archaea comes from the Greek word meaning "ancient." An appropriate name, because many archaea thrive in conditions mimicking those found more than 3.5 billion years ago. Back then, the earth was still covered by oceans that regularly reached the boiling point-an extreme condition not unlike the hydrothermal vents and sulfuric waters where archaea are found today.

Some scientists consider archaea living fossils that may provide hints about what the earliest life forms on Earth were like, and how life evolved on our planet.

In addition to superheated waters, archaea have been found in acid-laden streams around old mines, in frigid Antarctic ice and in the super-salty waters of the Dead Sea. A number of other extreme-living bacterial species also enjoy these conditions, too, such as the community of cyanobacteria and bacteria shown top right.

Foodborne illness or food poisoning is caused by consuming food contaminated with pathogenic bacteria, toxins, viruses, prions or parasites. Such contamination usually arises from improper handling, preparation or storage of food. Foodborne illness can also be caused by adding pesticides or medicines to food, or by accidentally consuming naturally poisonous substances like poisonous mushrooms or reef fish. Contact between food and pests, especially flies, rodents and cockroaches, is a further cause of contamination of food.

Some common diseases are occasionally foodborne mainly through the water vector, even though they are usually transmitted by other routes. These include infections caused by Shigella, Hepatitis A, and the parasites Giardia lamblia and Cryptosporidium parvum.

Thermophiles like unusually hot temperatures. A few species have been found to survive even above 110 degrees Celsius (water boils at 100 degrees Celsius). o Psychrophiles like extremely cold temperatures (even down to-10 degrees Celsius). o Halophiles thrive in unusually salty habitats. Some can thrive in water that's 9% salt; sea water contains only 0.9% salt. o Acidophiles prefer acidic conditions; Alkaliphiles prefer very alkaline environs.

Accumulating sufficient data on the behaviour of microorganisms in foods requires an extensive amount of work, and is costly. In addition, while data alone can describe the response of a microorganisms in food, they provides little insight into the relationship between physiological processes and growth or survival.

Scientists also use molecular tools to extract and compare bits of a particular kind of RNA from samples in order to determine if previously known or new microbes are present in a particular environment. This technique is widely used as a biomarker and for microbial ecology studies. It uses a particular kind of RNA known as 16S ribosomal RNA, or 16S rRNA.

Ribosomes are the gene-translating machines in all living things. When a gene on a piece of DNA is copied into a strand of messenger RNA and ferried out of the cell nucleus into the cell fluid, ribosomes there latch onto this mRNA. The ribosomes move along the mRNA strand, reading the code contained in its sequence of nucleotide bases (the As, Gs, Cs and Us, since U replaces T in RNA) and stringing the right amino acids together based on the code to build protein chains.

Pasteurization is typically associated with milk. There are two widely used methods to pasteurize milk: high temperature/short time (HTST), and ultra-high temperature (UHT). HTST is by far the most common method. Milk simply labelled "pasteurized" is usually treated with the HTST method, whereas milk labelled "ultra-pasteurized" must be treated with the UHT method. HTST involves holding the milk at a temperature of 161.5 degrees F (or 72 C) for at least 15 seconds. UHT involves holding the milk at a temperature of 280 degrees F (or 138 C) for at least two seconds.

Pasteurization methods are usually standardized and controlled by national food safety agencies (such as the USDA in the United States and the Food Standards Agency in the U.K.). These agencies require milk to be HTST pasteurized in order to qualify for the "pasteurized" label. There are different standards for different dairy products, depending on the fat content and the intended usage. For example, the pasteurization standards for cream differ from the standards for fluid milk, and the standards for pasteurizing cheese are designed to preserve the phosphatase enzyme, which aids in curing the cheese.

The HTST pasteurization standard was designed to achieve a 5-log (approximately one million-fold) reduction in the number of viable microorganisms in milk. This is considered adequate for destroying almost all yeasts, mold, and common spoilage bacteria and also to ensure adequate destruction of common pathogenic heat-resistant organisms (including particularly Mycobacterium tuberculosis, which causes tuberculosis and Coxiella burnetii, which causes Q fever). HTST pasteurization processes must be designed so that the milk is heated evenly, and no part of the milk is subject to a shorter time or a lower temperature.

HTST pasteurized milk typically has a refrigerated shelf life of two to three weeks, whereas ultra pasteurized milk can last much longer when refrigerated, sometimes two to three months. When UHT pasteurization is combined with sterile handling and container technology, it can even be stored unrefrigerated for long periods of time.

A BASIS FOR FOOD SAFETY

Because microorganisms are easily dispersed, display physiologic diversity, and tolerate extreme conditions, they are ubiquitous and may contaminate and grow in many food products. The behavior of microbial populations in foods (growth, survival, or death) is determined by the properties of the food (e.g., water activity and pH) and the storage conditions (e.g., temperature, relative humidity, and atmosphere). The effect of these properties can be predicted by mathematical models derived from quantitative studies on microbial populations.

Temperature abuse is a major factor contributing to foodborne disease; monitoring temperature history during food processing, distribution, and storage is a simple, effective means to reduce the incidence of food poisoning. Interpretation of temperature profiles by computer programs based on predictive models allows informed decisions on the shelf life and safety of foods.

In-or on-package temperature indicators require further development to accurately predict microbial behavior. We suggest a basis for a "universal" temperature indicator. This article emphasizes the need to combine kinetic and probability approaches to modeling and suggests a method to define the bacterial growth/ no growth interface. Advances in controlling foodborne pathogens depend on understanding the pathogens' physiologic responses to growth constraints, including constraints conferring increased survival capacity.

Ensuring the microbial safety and shelf life of foods depends on minimizing the initial level of microbial contamination, preventing or limiting the rate of microbial growth, or destroying microbial populations. With many foods, these strategies have been practiced successfully for thousands of years. However, in the last decade, the incidence of foodborne disease has increased in the industrialized world, despite the introduction of the Hazard Analysis and Critical Control Points (HACCP) concept and the promulgation of regulations in food safety. The increased incidence of foodborne disease is caused by changes in agricultural and food processing practices, increasing international trade in foods, and social changes (which include changed eating habits and increased population mobility).

This article develops the propositions that available quantitative information, properly applied, is a basis for improved food safety; the information available, largely an empiric description of microbial behavior in foods, highlights a lack of understanding of the physiology of foodborne pathogens; and knowledge of the physiology may lead to more precise control of foodborne bacteria and novel protocols to ensure the microbiologic safety of foods.

CHARACTERISTICS OF BACTERIA

Bacteria have inhabited the earth for approximately three and a half billion years and have colonized almost every conceivable habitat. In fact, the development of microbial populations is probably precluded only when liquid water is absent or conditions are so extreme that rapid denaturation of proteins outpaces their rate of replacement. The major characteristics that underpin the success of prokaryotes are small size and ease of dispersal, physiologic diversity, and tolerance of extreme conditions.

The temperature range over which microbial populations develop is-12°C (the temperature at which intracellular water freezes) to +112°C (the temperature at which liquid water is maintained only under elevated pressure). The pH range is pH 1 to pH 12, and the salinity range is zero to saturated.

Langeveld et al., who studied microbial development in biofilms in a tubular heat exchanger used to pasteurize milk, report the exploitation of different ecologic niches by bacteria. Through the ascending temperature range of the tube (~20°C to ~90°C), the dominant microbiota changed from gram-negative bacteria such as Acinetobacter, to coliforms to Streptococcus thermophilus to thermophilic bacilli. At the highest temperature, the wall of the exchanger was colonized by Thermus thermophilus. Thus, it appears that contaminants deposited along the length of the tube were selected by the in situ temperature, with the fastest-growing organism dominating.

FACTORS AFFECTING MICROBIAL BEHAVIOR IN FOODS

Most studies in food microbiology are concerned with the rapid growth of populations, but in many ecosystems, the survival characteristics of the population also need to be considered. The longevity of bacterial spores and their resistance to harsh conditions are well documented. However, the ability of vegetative cells to resist stressful conditions is increasingly recognized as an important ecologic trait. Attention also needs to be given to relatively slow-growing populations in various situations, e.g., when the shelf life of a product is extended by control of rapidly growing spoilage organisms.

The behavior of foodborne microorganisms, be it the growth or death of microbial populations, is based on the time of exposure to environmental factors affecting population development; for example, equivalent kills of bacteria in milk are achieved by low temperature-long time pasteurization (60°C/30 min) and high temperature-short time pasteurization (72°C/15 sec).

When populations are in the biokinetic range, the rate at which they develop is determined by factors such as temperature, water availability, and pH applied in food preservation procedures. The extent of microbial growth is a function of the time the population is exposed to combinations of intrinsic food properties (e.g., salt concentration and acidity) and extrinsic storage conditions (e.g., temperature, relative humidity, and gaseous atmosphere).

Different factors assume dominance in different foods and preservation strategies. In many foods, the full preservation potential of a single property is restricted because of considerations related to the esthetic, organoleptic, and nutritional properties of the product. However, several properties or conditions may be combined to provide a desired level of stability. In situations where the preservation strategy is designed to slow the rate of population growth, the effect will always be increased by storage temperature.

Temperature control in processing, distribution, and storage (the cold chain) is crucial to ensure the adequate shelf life and safety of many common foods, including meat, fish, poultry, and milk. Newer technologies, including modified atmosphere packaging and sophisticated products such as sous-vide meals, do not obviate the need for strict temperature control. Indeed, the requirement for vigilance increases with increased shelf life and the possibility of growth of psychrotrophic pathogens over an extended period.

PREDICTIVE MICROBIOLOGY

Predictive microbiology involves knowledge of microbial growth responses to environmental factors summarized as equations or mathematical models. The raw data and models may be stored in a database from which the information can be retrieved and used to interpret the effect of processing and distribution practices on microbial proliferation. Coupled with information on environmental history during processing and storage, predictive microbiology provides precision in making decisions on the microbiologic safety and quality of foods. The term "quantitative microbial ecology" has been suggested as an alternative to "predictive microbiology."

The development, validation, and application of predictive microbiology has been extensively reviewed in the last decade. Modeling studies have concentrated on descriptions of the effect of constraints on microbial growth (rather than survival or death), often using a kinetic model approach (rather than probability modeling) and most often describing the effect of temperature as the sole or one of a number of controlling factors. For example, the temperature dependence model for growth of Clostridium botulinum demonstrated a good fit to data, but the authors noted that "care must be taken at extremes of growth, as no growth may be registered in a situation where growth is indeed possible but has a low probability."

The emphasis in modeling efforts on temperature (often in combination with other factors) may be justified, given its crucial role in the safe distribution and storage of foods. Surveys carried out over several decades in the United Kingdom, United States, Canada, and Australia point to the predominant role of temperature abuse in outbreaks of foodborne disease.

PROBLEMS WITH PREDICTIVE MICROBIOLOGY AND RESEARCH NEEDS

Several commonly perceived problems with predictive modeling are reviewed below. While considerable progress has been made in defining philosophic approaches and experimental protocols for growth model development and many models have been developed and published, more validation studies are required, particularly involving independent and industry-based trials. More emphasis should be placed on modeling the death kinetics of foodborne pathogens with low infective doses.

Measurement of environmental factors (e.g., temperature) can be achieved with precision, but in some situations, (e.g., in chilling of meat carcasses), it is more difficult. Location of the sensor can be an important consideration. In meat chilling, where control of microbial development is a function of the combined effects of falling temperature and water activity, development of a technique to measure water activity in situ at the carcass surface would provide valuable information.

Furthermore, development of techniques to measure constraints such as water activity, pH, or redox potential on a microscale might provide useful information for a complex food such as salami. This would allow definition of the role of the microenvironment in determining microbial behavior. The concept of a microenvironment is well developed in soil microbiology but has been neglected in food microbiology.

The inherent variability of response times (generation time and lag phase duration) as an issue in predictive microbiology was first raised by Ratkowsky et al., who related the variance of responses on a transformed rate scale such as V(Ök) or V(lnk) to the variances of responses on a time (q) scale. The variance was shown to be proportional to the square or cube of the response time. It was later confirmed that nonnormal gamma and inverse Gaussian distributions described the distribution of response times in predictive microbiology, which indicate that variance is proportional to the square or cube of the response time, respectively.

The practical implication of these findings for the application of kinetic models is that inherent biologic variability increases markedly with increasing response times, and thus the confidence limits associated with predictions also increase markedly. However, if the probability distribution of the response time is known, one can determine the probability that an organism will grow more quickly than a predicted response time. Thus, kinetic models are appropriate to describe consistent microbial growth responses, but under extreme conditions a probability approach may be required.

Models are normally developed under static conditions (growth rates and lag times are measured at a series of set temperatures, water activity values, and pH levels), and the results are combined to describe the effects of each factor or a combination of factors on population development. Subsequently, models must be validated in foods under conditions that mimic situations encountered in normal practice, e.g., decreasing temperature and water activity during active chilling of meat carcasses or fluctuating temperatures during the distribution and storage of many food commodities.

Shaw and later several other authors reported on the effect of fluctuating temperatures. Depending on the magnitude of the temperature deviation, the organism may change its rate of growth to a rate characteristic of the new temperature, or it may stop growing if a lag phase is introduced. In both situations, Salmonella Typhimurium has responded nearly as predicted by the model. Baranyi et al. presented similar results for the spoilage bacterium Brocothrix thermosphacta. When cycled between 25°C and 5°C, the model predicted behavior well in both the growth rate and lag phase duration.

However, a temperature shift from 25°C to 3°C caused deviations from model predictions due to decline in viable cell numbers or extended lag phases. During the final extended phase of growth at 2.8°C, the rate resumed at that predicted. Baranyi et al. also examined the perceived problem of modeling lag phase duration. The difficulty in predicting lag phase duration in foods is not due to the lack of a suitable model: the difficulty comes from the lack of knowledge of the physiologic status of the organisms contaminating the food. The organisms may include cells that are actively growing, exhibiting a physiologic lag phase, damaged and under repair, exhibiting physiologic (endospores) or exogenous dormancy (VNC cells), damaged but unable to reproduce because of ineffective repair mechanisms, and dead. At least part of the confusion surrounding the measurement of lag phase duration arises from experiments in which inocula of different physiologic statuses were used.

Methods to define the physiologic status of foodborne contaminants under various conditions need to be developed. This will require observations on individual cells or small populations of cells either directly by microscopy or an indicator of single-cell metabolic activity. Luminescent Salmonella strains have been used as real-time reporters of growth and recovery from sublethal injury. Alternatively, a parameter to describe the suitability of cells to grow in a new environment may be incorporated in the model.

CURRENT STATUS OF PREDICTIVE MICROBIOLOGY

Some problems with predictive microbiology have been discussed, and, for each, needed research has been suggested.

Opinions vary on the efficacy of models to predict outcomes under real life conditions. At one end of the scale, accuracy such as that described for the growth of Pseudomonas in minced beef (Figure 1) can be obtained in trials conducted independently of the laboratory in which the model was developed. At the other end, models developed in laboratory broth systems have been reported to be inappropriate to describe growth on food.

Dalgaard reported similar discrepancies and suggested an iterative approach to model development using food, rather than laboratory media, as the growth substrate for model development. Such reports emphasize the need for rigorous validation of models under practical conditions. Deviations from predictions do not necessarily imply that the model is defective but more likely that knowledge of some food ecosystems is incomplete and factors other than those used in model development have an effect on microbial behavior.

The common theme of the problems in predictive microbiology discussed above is that of uncertainty-uncertainty in terms of the starting conditions (e.g., initial microbial numbers and types) and the microbial response in a given or changing environment. Uncertainty translates to variability if the distribution of response times is understood and the variance can be described. As we have indicated above, the variability associated with very long response times limits the utility of kinetic models and requires a probability approach.

Thus, while in the last decade predictive modelers were justified in their selection of temperature as a primary factor to model in kinetic approaches, the next decade may see a return to probability modeling as pioneered by Genigeorgis and Roberts. This shift will derive impetus from the emergence of dangerous pathogens with very low infective doses, and continued kinetic modeling will concentrate on survival and death rather than growth of populations.

The first kinetic death model to find widespread use in the food industry was for thermal destruction. One can consider a model describing a 12-log cycle reduction in C. botulinum spores in a short time with considerable certainty. However, as we move toward less severe processes with longer response times and the added complications of "shoulders" and "tails" to define the growth/no growth interface, biologic variability will again dictate a probability approach to describe the survival and slow decline of microbial populations.

DEFINING THE GROWTH/NO GROWTH INTERFACE

Because growth of pathogenic bacteria in foods always increases the risk for foodborne disease, defining the conditions at which no growth is possible is of considerable practical significance for food manufacturers and regulators.

Bacterial growth/no growth interface models quantify the combined effect of various hurdles on the probability of growth and define combinations at which the growth rate is zero. Increasing the level of one or more hurdles at the interface by only a small amount will significantly increase the probability of "fail safe" events and decrease the probability that a few cells in the population will resolve the lag phase and begin to grow (a "fail dangerous" event).

The growth/no growth interface also has great physiologic significance because at that point biosynthetic processes are insufficient to support population growth, and survival mechanisms are in place.

A procedure to derive the interface was proposed by Ratkowsky and Ross; it employs a logistic regression model to define the probability of growth as a function of one or more controling environmental factors. From this model, the boundary between growth and no growth, at some chosen level of probability, can be determined.

The form of the expression containing the growth limiting factors is suggested by a kinetic model, while the response at a given combination of factors is either presence or absence (i.e., growth/no growth) or probabilistic (i.e., the fraction of positive responses in n trials). This approach represents an integration of probability and kinetic approaches to predictive modeling.

MICROBIAL RESPONSES TO STRESS AND MICROBIAL PHYSIOLOGY

Bacteria have physiologic mechanisms enabling them to survive in environments that preclude their growth. While some tolerance to environmental insults is adaptive, a wide range of protective mechanisms is induced when cells enter a stationary phase or become starved.

These phenomena are under the control of the rpoS gene, which codes for a stationary-phase-specific sigma factor, expression of which triggers the development of a semidormant state in which bacteria can better resist multiple physical challenges. This factor and the gene products whose expression it controls are of vital significance to food microbiology; they form the basis for a global stress response in which one stress can confer protection to a wide range of other stresses. Under the influence of this factor, bacterial cells respond very quickly to unfavorable changes in their environment. Often these responses are phenotypic and remain in place only during stress.

LOW PH TOLERANCE

Brown et al. demonstrated "acid habituation," a phenotypic response to an environmental insult, for five strains of Escherichia coli. These strains showed a wide range of intrinsic acid tolerance, which for each strain was enhanced by exposure to nonlethal acidity (pH 5) before exposure to a lethal acid challenge (pH 3). Neutralization of the growth medium partially reversed tolerance to acid stress, underlining that acid habituation is a phenotypic response.

Furthermore, acid tolerance was correlated with changes in the fatty acid composition of the cell membrane. During acid habituation, monounsaturated fatty acids (16:1w7c and 18:1w7c) in the phospholipids of E. coli were either converted to their cyclopropane derivatives (cy17:0 and cy19:0) or replaced by saturated fatty acids. The degree of acid tolerance of the five strains of E. coli was highly correlated with the membrane cyclopropane fatty acid content, which may enhance the survival of cells exposed to low pH.

LOW WATER ACTIVITY TOLERANCE

Bacterial cells, when confronted by lowered water activity, regulate the internal environment by rapidly accumulating compatible solutes such as glycine betaine or carnitine. The solutes, which may be scavenged by the cell, exert their influence at very low concentrations; the effect is demonstrated both in limits and rate growth. These compounds appear also to provide cryotolerance as well as osmotolerance.

ENERGY DIVERSION

Microbial responses to stressful conditions may constitute a drain on the energy resources of the cell, e.g., in relation to the accumulation of compatible solutes. Knochel and Gould argued "that restriction of the availability of energy will interfere with a cell's reaction to osmotic stress." The energy diversion hypothesis was supported by McMeekin et al. on the basis of observations on the growth of Staphylococcus xylosus at nine different levels of water activity. Though Tmin, the theoretic minimum temperature for growth, remained constant, the actual minimum temperature at which growth was observed increased with decreasing water activity, suggesting energy expenditure to cope with aw stress.

Krist and Ross, however, challenged this explanation because of findings from growth rate and yield experiments on E. coli growing in a glucose-limited minimal minerals medium. With both decreasing temperature and water activity, the growth rate declined gradually, but the yield was not greatly affected until close to the point where growth ceased. As the substrate was converted to the same amount of biomass, this suggests that the stresses imposed by suboptimal temperature or water activity are not a major drain on the cell's energy reserves.

Compatible solutes likely ameliorate the effect of both factors by maintaining enzymes in an active configuration. With pH, the growth rate of many organisms is unaffected across a wide range of pH values. To maintain intracellular pH, the cell uses considerable energy to export protons, and thus it is anticipated that yield will be affected.

Increasing knowledge of the physiologic response of bacterial cells to individual constraints and combinations of constraints will provide greater precision in defining growth-limiting conditions and possibly allow development of novel protocols to ensure the microbial safety of foods. As an example, the remarkable effect of compatible solutes on the growth rate and growth range conditions is an obvious advantage for bacteria growing under stressful conditions (Figure 2). Compatible solutes, such as betaine and carnitine, are widely distributed in foods of plant and animal origin and are easily available to bacteria and rapidly taken up by specific transport mechanisms. It is unlikely that growth might be controlled by "creating a hostile environment devoid of osmolytes," as suggested by Smith.

However, it might be possible to use the specific uptake mechanism to deliver a compatible solute analogue with lethal effects on the cell. Alternatively, if the cell is moved from an environment in which growth is possible to one where growth ceases, compatible solutes may also allow enzymic reactions to continue within the cell, depleting energy reserves and inducing a greatly extended lag phase or death. This hypothesis is supported by the observations of many authors that survival is better at low rather than ambient temperatures.

For example, Clavero and Beuchat state, "Regardless of the pH and aw, survival of E. coli O157:H7 was better at 5°C than at 20°C or 30°C." Furthermore, preliminary experiments in this laboratory suggest that E. coli declines more rapidly in the presence of betaine than in its absence.

APPLICATION OF PREDICTIVE MODELS

The incorporation of predictive models into devices such as temperature loggers has been described for E. coli and Pseudomonas, as has the development of expert systems from predictive modeling databases.

The current food poisoning crisis indicates that existing quantitative information on microbial growth, survival, and death, if properly applied, would have an immediate impact on the incidence of foodborne disease in the industrialized world. Even without the synthesis of data into mathematical models, simply logging the temperature history of food processing, distribution, and storage operations would provide much useful information. Loggers provide a hard copy of a temperature profile in real time and thus evidence of temperature abuse and the source of the abuse.

For loggers with appropriate software, the temperature profile may be interpreted in terms of microbial growth. However, the interpretation must be based on an informed analysis of the temperature history by a trained operator. The operator may, for example, be required to enter default values for initial bacterial numbers or provide an estimate of lag phase duration under specified conditions. Estimates of both imply general knowledge of food microbiology and specific knowledge of the process and products under consideration. The equivalence of an estimate of microbial growth derived from temperature profile to that obtained from conventional microbiologic criteria may also be necessary.

An alternative to the use of temperature loggers is the development of in- or on-package temperature tags as recommended in the U.S. Food Safety Initiative draft document Food Safety from Farm to Table. With temperature tags, informed interpretation is not required because abuse is indicated directly by the tag response. Therefore, the tag must indicate the significance of the environmental history for microbial behavior. The time/temperature tags available are based on physical or chemical changes that follow Arrhenius kinetics. While these may give a reasonable approximation of microbial growth in the normal range, the deviation of microbial responses becomes increasingly large as conditions move from normal to stressful. The intriguing possibility of a universal time/temperature indicator was flagged on the basis of observations made of temperature effects on foodborne pathogens in this laboratory and by Snyder. The universal indicator is based on a relationship that describes the maximum specific growth rate of a continuum of organisms from psychrophiles to thermophiles in terms of Arrhenius kinetics with an apparent activation energy of ~80 kJ/mole. This value can be related to the activation energy/growth rate at any other temperature by a relative rate function derived from Belehradek (square root) kinetics.

CONCLUSIONS

We have argued that a thorough understanding of microbial ecology and physiology offers the best opportunity to control microbial populations in food and reverse the upward trend in the incidence of foodborne disease. Many food preservation strategies have their origin in empirical observations practiced for thousands of years. The systematic collection and collation of data on microbial behavior in foods spawned the discipline of food microbiology, within which predictive microbiology (quantitative microbial ecology) has accelerated our understanding of the microbial ecology of foodborne bacteria. Studies in microbial physiology will further enhance our knowledge and offer new possibilities for food preservation.

The most disturbing aspect of the current crisis is that simple application of existing knowledge would lead to a marked reduction in the incidence of foodborne disease. Education of food handlers and consumers in basic hygiene and the consequences of temperature abuse is urgently needed as is a greater depth of understanding for those in technical positions in the food industry or those with regulatory responsibilities. Furthermore, an appreciation of the need for shared responsibility for food safety within all sectors of the continuum from farm to table, including the consumer, has to be developed. The U.S. Food Safety Initiative draft document emphasizes this point, as does the structure of the Australian Meat Research Corporation's Microbial Food Safety Key Program.

5

FOOD PROCESSING AND BIOTECHNOLOGY

Food processing is the set of methods and techniques used to transform raw ingredients into food for consumption by humans or animals. The food processing industry utilises these processes. Food processing often takes clean, harvested or slaughtered and butchered components and uses these to produce attractive and marketable food products. Similar process are used to produce animal feed.

Examples

Following are common food processing techniques:

- removal of unwanted outer layers, such as potato peeling or the skinning of Peaches
- Chopping or slicing, of which examples include potato chips, diced carrot (Brent), or candied peel.
- Mincing and macerating
- Liquefaction, such as to produce fruit juice
- Emulsification
- Cooking, such as boiling, broiling, frying, steaming or grilling
- deep frying
- Mixing
- Addition of gas such as air entrainment for bread or gasification of soft drinks

- Proofing
- spray drying.

Extreme examples of food processing include the delicate preparation of deadly fugu fish, preparing space food for consumption under zero gravity, winemaking, hot dogs,Mauritz, Niel and chicken nuggets.

HISTORY

Food processing dates back to the prehistoric ages when crude processing incorporated slaughtering, fermenting, sun drying, preserving with salt, and various types of cooking (such as roasting, smoking, steaming, and oven baking). Salt-preservation was especially common for foods that constituted warrior and sailors' diets, up until the introduction of canning methods. These crude processing techniques remained essentially the same until the advent of the industrial revolution.

Modern food processing technology in the 19th and 20th century was largely developed to serve military needs. In 1809 Nicolas Appert invented a vacuum bottling technique that would supply food for French troops, and this contributed to the development of tinning and then canning by Peter Durand in 1810.

Although initially expensive and somewhat hazardous due to the lead used in cans, canned goods would later become a staple around the world. Pasteurization, discovered by Louis Pasteur in 1862, was a significant advance in ensuring the micro-biological safety of food.

In the 20th century, World War II, the space race and the rising consumer society in developed countries (including the United States) contributed to the growth of food processing with such advances as spray drying, juice concentrates, freeze drying and the introduction of artificial sweeteners, colorants, and preservatives such as sodium benzoate and saccharine. In the late 20th century products such as dried instant soups, reconstituted fruits and juices, and self cooking meals such as MRE food ration were developed.

Because the 20th century witnessed a rise in the pursuit of convenience, food processors especially marketed their products to middle-class working wives and mothers. Frozen foods (often credited to Clarence Birdseye) found their success in sales of juice concentrates and Swanson's "TV dinners". Processors utilized the perceived value of time to appeal to the postwar population, and this same appeal contributes to the success of convenience foods today.

BENEFITS

Benefits of food processing includes toxin removal, preservation, improving flavor, easing marketing and distribution tasks, and increasing food consistency. In addition, it increases seasonal availability of many foods, enables transportation of delicate perishable foods across long distances, and makes many kinds of foods safe to eat by removing the microorganisms. Modern supermarkets would not be feasible without modern food processing techniques, long voyages would not be possible, and military campaigns would be significantly more difficult and costly to execute.

Modern food processing also improves the quality of life for allergics, diabetics, and other people who cannot consume some common food elements. Food processing can also add extra nutrients.

DRAWBACKS

Food processing can lower the nutritional value of some foods. Some preservatives added or created during processing such as nitrites or sulfites may cause adverse health effects on some consumers. In addition, high quality and hygiene standards must be maintained to ensure consumer safety and failures to maintain adequate standards can have serious health consequences.

In general, fresh food that has not been processed other than by washing and simple kitchen preparation, may be expected to contain a higher proportion of naturally occurring vitamins, fibre and minerals than the equivalent product processed by the food industry. However fresh materials are more liable to early spoilage and are often unsuited to long distance transportation from source to shelf. Fresh materials, such as fresh produce and raw meats, are also more likely to harbour pathogenic microorganisms (e.g. Salmonella) capable of causing serious illnesses.

INDUSTRIES

Food processing industries and practices include the following:

- Meat packing plant
- Industrial rendering
- Slaughterhouse
- Vegetable packing plant
- Cannery.

FOOD ALLERGY AND INTOLERANCE

Food allergy is often mistaken for food intolerance. It is important to note that allergy is only one of a number of possible reasons for food intolerance. Food intolerance can be defined as a condition where particular adverse effects occur after eating a particular food or food ingredient. Genuine food intolerance is different from psychologically based food aversion, where a person strongly dislikes a food and believes that a food produces a particular reaction.

A genuine food allergy is when a specific immune reaction occurs in the body in response to consuming a particular food. Allergies often run in families, and people who are allergic to some foods may also be allergic to other environmental factors, such as house dust, animal fur and pollen.

A true allergic response involves an altered or abnormal tissue reaction to an antigen. An antigen can be a protein, a substance bound to a protein, a food additive or less commonly, a polysaccharide. The antigen combines with an antibody and produces an immune response, which results in cell damage and the release of histamine.

The immune system plays an essential role in our bodies in protecting us from the invasion of harmful substances. An allergy occurs when the mechanism operates inappropriately in response to a harmless substance such as a particular food protein. Food intolerances, other than allergies, can occur for a variety of reasons including:

NON ALLERGIC HISTAMINE RELEASE

The signs are very similar to an allergy and include headache, swelling, urticaria, vomiting and diarrhoea. A substance called histamine is released (it is also released in true allergic reactions) in response to foods such as shellfish or strawberries.

METABOLIC DEFECTS

A lack or deficiency of enzymes responsible for the digestion of food can cause many types of food intolerance. For example, a deficiency in lactase, the enzyme responsible for digesting milk, causes intolerance to milk. Coeliac disease is a gut intolerance to a protein found in wheat, called gluten, it would not be considered an allergy. The symptoms of coeliac disease are controlled by following a gluten-free diet. It is unknown exactly why or how gluten harms the gut, although it is now thought to be an abnormal immunological response rather than an enzyme deficiency. It is still not considered to be a food allergy in the true sense of the definition.

PHARMACOLOGICAL EFFECTS

Some food substances can act like drugs, particularly if taken in large quantities. The most familiar of these substances is caffeine, found in tea, coffee, chocolate and cola drinks. A large intake of caffeine can cause tremor, migraine and palpitations. Other pharmacologically active substances found in food include histamine, tyramine, tryptamine and serotonin, which may be consumed in foods such as red wine, cheese, yeast extract, avocados and bananas. In susceptible people, these foods can trigger urticaria, facial flushing and headaches.

FOOD INTOLERANCE OF UNKNOWN ORIGIN

Reactions can be provoked by many foods and food products which we cannot be clear about. They may or may not be allergic reactions. Food additives, particularly tartrazine and sodium benzoate, can provoke urticaria, rhinitis and asthma. Yeasts can provoke a number of reactions in some people, particularly skin disorders.

COMMON CAUSES OF FOOD INTOLERANCE

The most common food intolerances, in order of frequency are milk, eggs, nuts, fish/shellfish, wheat/flour, chocolate, artificial colours, pork/bacon, chicken, tomato, soft fruit, cheese and yeast. Whilst not all food intolerances are related to meat and dairy products, it can be seen from the above list that vegetarians, and particularly vegans, will suffer less from food intolerance because they already eliminate some of the most common causes of intolerance.

SYMPTOMS

The most common symptoms of allergy include asthma, gastro-intestinal symptoms (nausea, vomiting, and diarhhoea), eczema, urticaria (hives), rhinorrhea (heavy discharge from the nose), and angio-oedema (swelling of the blood vessels). Other more long-term symptoms include can depression, anxiety, fatigue, migraine, sleeplessness and hyperactivity in children.

TREATMENT

As it is sometimes quite difficult to distinguish between a genuine food allergy and a food intolerance, treatment is often similar. The first step is to diagnose the food intolerance. This should not be done without medical supervision as some reactions to food intolerance can be dangerous. Sometimes the cause of a particular food intolerance is obvious, by the immediate effect that occurs on eating a particular food. In this case the treatment is simply to avoid that particular food. In most cases the suspected food is more difficult to track down. A diary kept of foods eaten and symptoms experienced can sometimes help detect the offending food or foods. Other factors such as the weather, menstrual cycles and difficult relationships can affect the symptoms. Sometimes simple exclusion diets are advised where record keeping suggests a particular food may be the cause. So, for example, milk, egg or wheat may be avoided to see if symptoms improve.

Other more restrictive diets may be advised, which only include a limited amount of foods which rarely cause a reaction. These diets are usually called exclusion diets. The idea of an exclusion diet is to identify an allergy or intolerance, by limiting the food to a very small choice, checking for symptoms and then very gradually introducing test foods to see if there is a reaction.

An exclusion diet should not be followed without sound nutritional advice. You can seek the help and advice of The Vegetarian Society if you have any problems with your vegetarian or vegan diet, relating to diagnosis or treatment of a food allergy or intolerance.

DEFINITIONS

A vegetarian is someone living on a diet of grains, pulses, nuts, seeds, vegetables and fruits with or without the use of dairy products and eggs. A vegetarian does not eat any meat, poultry, game, fish, shellfish or crustacea, or slaughter by-products.

TYPES OF VEGETARIAN

Lacto-ovo-vegetarian: Eats both dairy products and eggs. This is the most common type of vegetarian diet.

Lacto-vegetarian: Eats dairy products but not eggs.

Vegan: Does not eat dairy products, eggs, or any other animal product.

STUMBLING BLOCKS

Many foods contain ingredients derived from the slaughter of animals. Gelatine is made from animal ligaments, tendons, bones etc which have been boiled in water. It is often found in confectionery, low fat spreads and desserts, and other dairy products. The term animal fat refers to carcass fat and may be present in a wide range of foods, including biscuits, cakes, and margarines. Suet and lard are types of animal fats. Certain food additives (E numbers) may be derived from animal sources. Cheese is often made with rennet extracted from the stomach lining of slaughtered calves. Vegetarian cheese is made with rennet from a microbial source. The Vegetarian Society has an information sheet listing ingredients which may be unsuitable for vegetarians. Many vegetarians that eat eggs will eat only free-range eggs. This is due to moral objections to the battery farming of hens. The Vegetarian Society only endorses products containing eggs if the eggs are certified as free-range.

PRACTICES IN CATERING

Vegetarians dining out will expect work surfaces and chopping boards, utensils and all other kitchen equipment and facilities to be either kept separate from those used for non-vegetarian food preparation, or cleaned thoroughly before vegetarian food preparation. Caterers should also ensure that fryers, grills and griddles used for preparing non-vegetarian products are thoroughly cleaned. Fryers must be filled with fresh, uncontaminated oil before vegetarian food is cooked. The Society recommend that caterers keep a separate set of utensils for the preparation and serving of vegetarian meals.

GLUTEN-FREE DIET

A gluten-free diet is essential for people who have coeliac disease or dermatitis herpetiformis (a gluten induced skin sensitivity). Some people may choose to follow a gluten-free diet for other reasons, although these two diseases are the only ones where a gluten-free diet is considered medically imperative. Gluten is a mixture of proteins found in some cereals, particularly wheat. It is the gliadin component of gluten which is responsible for coeliac disease.

A gluten-free diet is not the same as a wheat-free diet, and some gluten-free foods are not wheat free. Despite a good deal of research, it is unknown how or exactly why gluten harms the gut. It is now considered likely that coeliac disease involves an abnormal immunologic response, rather than an enzyme deficiency as was suggested in the past.

The Vegetarian Society believes that a gluten-free diet is compatible with vegetarianism. Some doctors and The Coeliac Society advise against a vegetarian or vegan diet for coeliacs because they believe it may make your diet too complicated and this could mean it is difficult to comply with. There are no known medical or nutritional reasons why you should not be a vegetarian or vegan coeliac, although the gluten-free aspects of your diet must be the priority for your own health and well-being. It is possible to follow a gluten-free vegan diet, although you must be extra careful to ensure that your diet is nutritionally adequate. It is essential that you seek the advice of a sympathetic dietitian if you want to follow a vegan gluten-free diet. Vegetarians may initially find it difficult to establish what foods they can and cannot have. This Information Sheet is designed to help.

A gluten-free diet involves the complete avoidance of all foods made from or containing wheat, rye, barley and usually, oats. Some doctors say oats may be permitted, although The Coeliac Society advise against the inclusion of oats in a gluten-free diet. The Coeliac Society publishes a list of gluten-free manufactured products in a booklet which is updated every year. You can check with The Vegetarian Society if you are unsure whether any particular foods on this list are suitable for vegetarians or vegans. Some manufacturers use the gluten-free symbol on their label. A wide range of specially manufactured gluten-free foods such as, bread, bread mix, pasta, biscuits, cakes, crispbread and flour are prescribable under the NHS. Some groups of people are exempt from prescription charges, children, pregnant women and pensioners in particular. If you are not exempt, it works out economical to buy a "season ticket" type prescription. Some gluten-free products, such as chocolate biscuits, are considered luxuries and are not prescribable, although they can be bought from the chemist.

Nutritional Advice

Coeliac disease leads to severe damage of the gut surface, which can be completely reversed by following a gluten-free diet. Shortly after the diagnosis of coeliac disease, you need to be extra careful to ensure you have a nutritionally adequate diet, as you may have been suffering from malabsorption of nutrients.

Protein

Some gluten-free flours are low in protein, because they have had the gluten removed, which is itself a protein. Specially manufactured, prescribed gluten-free flours usually have milk protein added. Vegetarians can get protein from nuts & seeds, pulses, the non-gluten containing cereals, soya products, milk, cheese and free range eggs. Make sure some protein is included in each meal, and practice protein complementation with the vegetable proteins, for example, combine a nut or pulse dish with a suitable cereal. Protein is especially important to a growing child, it is essential to seek the advice of your dietitian if you are bringing up a child on a gluten-free vegetarian or vegan diet.

Anaemia

Following diagnosis, many coeliacs sufferers are anaemic. This is usually due to iron deficiency, although it could also be due to folic acid or vitamin B12 deficiency. Your doctor may prescribe an iron supplement for you until your digestive system is back to normal and can absorb iron again. To ensure a good intake of iron include pulses, lentils, nuts and green vegetables daily in your diet.

Avoid drinking tea with meals and instead have fruit juice, which helps the absorption of iron because it contains vitamin C.

Going Vegetarian

If you are already a vegetarian or vegan and are advised to follow a gluten-free diet, you do not need to abandon your vegetarianism. If you are recently diagnosed and would like to become vegetarian or vegan, do give your digestive system time to recover before making major changes to your diet, after all, you have a lot to think about in getting used to a gluten-free diet. When you have stabilised, you can gradually change to a vegetarian diet. Do contact The Vegetarian Society if you have any problems, when making the change to a vegetarian way of living. Also, do contact us if you have any difficulties with health professionals, who are sceptical about your following a vegetarian diet. Once coeliac disease has been diagnosed, it is recommended that you follow a gluten-free diet for life.

Vegetarian Gluten-Free and Gluten Containing Foods		
Food	**Gluten Free**	**Gluten Containing**
Cereals, flours, cakes & biscuits	Arrowroot, buckwheat, corn/maize, potato flour, rice, rice bran, rice flour, sago, tapioca, soya, soya bran, soya flour.	Wheat, wholemeal, wholewheat & wheatmeal flour, wheat bran, barley, rye, rye flour, pasta, noodles, semolina, All baked foods made from wheat, rye, semolina, barley & pearl barley.
Dairy products & eggs	Eggs, milk, cream, butter, cheese, curd cheese, coffee whiteners, soya products	Some yoghurts (eg muesli yoghurt), some cheese spreads
Puddings	Tapioca, sago, rice, custard.	Semolina, sponge pudding, pastry, pies, wafers.
Beverages	Tea, coffee, herb tea, fizzy drinks, fruit squash, cocoa, marmite, most alcoholic drinks.	Barley based drinks, barley fruit drinks, malted drinks, beer.
Fruit & vegetables	All fruits & vegetables, unless with certain sauces or cooked with gluten containing foods	.
Nuts, seeds & pulses	All, except certain brands of baked beans and beans with a gluten containing sauce	.
Preserves & confectionary	Jam, marmalade, sugar, honey, treacle, molasses, golden syrup, some brands of chocolate & sweets.	Confectionary containing flour, some brands of lemon curd
Soups & sauces	Gluten-free if thickened with a suitable flour. Many manufactured sauces, stock cubes, and soups contain gluten	.
Miscellaneous	Salt, pepper, vinegar, herbs & spices, tamari, yeast, most food colourings and essences.	Some pepper compounds, shoyu, ready-mix spices, some seasoning powders, certain brands of mustard. Certain medicines may contain gluten.

INFANT DIET

Weaning Vegetarian Babies

Taking the first steps in bringing your child up as a vegetarian isn't difficult. Remember that the nutritional requirements of a small baby are high, needing proportionally more protein, calcium and most other nutrients than at any other time of life. It is now widely recognised by bodies such as the British Medical Association, that a vegetarian diet can provide all the nutrients needed for growing infants.

Bringing up your child as a vegetarian, you will want to get them used to the vegetarian food groups: cereals, beans, nuts and seeds, dairy and soya produce, fruit and vegetables. Your baby may reject stronger-tasting foods, such as broccoli, cauliflower and cabbage, at six months but may like them several months later. It is quite safe to bring up your baby as a vegan, with no animal foods at all, as long as you make sure that plenty of nutrient-rich foods are included. Vegan babies need good sources of calcium, vitamin B12, vitamin D and protein.

Weaning is a gradual process that begins when you start to replace milk with solid foods. Infants should not be given solid foods before the age of four months except on the advise of a health professional.

A mixed diet should be offered by the age of six months, as soon thereafter babies will need a source of iron in their diet as breast or formula milk can no longer provide enough. Especially if there is a family history of allergies, when you begin weaning your baby, introduce one food at a time and leave a few days between each new food. This way, you will be able to tell if your baby is allergic or sensitive to any particular food.

Stages of Weaning

Before 6 Months

Before the age of four months babies can't properly digest any foods other than breast or formula milk, which remains by far the most important source of nutrition beyond the age of six months. Current recommendations are not to begin the weaning process before the age of six months, however babies can be introduced to solids before the age of six months, particularly if they seem hungry after their breast feed, on the advice of a Health Professional. Do not give wheat, oats, milk, nuts, eggs or citrus fruit if you decide to start weaning before the age of six months.

6-7 Months

Start by introducing one teaspoon of pureed fruit or baby rice mixed with breast or formula milk after a milk feed or in the middle if this works best for your baby. Take care that the food is adequately cooled. The nutrition of the food is not so important initially as milk still supplies all the baby's needs.

Other foods to try initially are:

- o Puréed vegetables, such as potatoes, carrot or courgette
- o Puréed fruit, such as banana, cooked apple or pear
- o Baby rice, corn meal, sago or millet can all be given as a thin porridge.

As the weeks go on milk remains the most important food in your baby's diet, but you can gradually increase the number of times solid food is given from once to twice a day. Try mashed lentils with some added vegetable oil and a wider variety of fruit and vegetables such as avocado and green vegetables.

7-8 Months

Most babies will by now be regularly eating solids although milk is a large part in their diets. In particular solid foods now provide an important source of iron. You can puree or sieve family foods to give variety, as long as they do not contain added salt. Try introducing tofu and mashed lentils if your baby is not already eating them. Dairy foods (yoghurt and cheese) can be introduced after six months, as can eggs as long as they are hard-boiled, though some experts recommend avoiding all dairy products and eggs until 12 months old.

8-12 Months

Your baby will gradually be able to cope with lumpier foods. Foods from the family table can be given as long as they do not contain salt. Well cooked and mashed peas and beans can be introduced from around 8 months. They are difficult to digest and so can cause problems if introduced earlier. Avoid sweet biscuits and rusks. Try introducing pieces of peeled apple, raw carrot or crusts of bread. When your baby is able to chew pieces of fruit, sandwiches and toast can become normal everyday foods. By the age of 12 months your baby should be enjoying three meals a day. Many companies produce baby foods suitable for vegetarians but it's quick and easy to prepare your own food for your baby. If you buy shop bought food, always check the ingredients label and look out for The Vegetarian Society's Seedling symbol to be absolutely sure that it is totally vegetarian.

Important Nutrients

Iron

This is an important nutrient during weaning, as milk is a very poor source of iron. Babies are born with their own store of iron but this will be depleted by six months. Although iron is less easily absorbed from non-animal sources, there are plenty of good vegetable sources. Iron-rich foods suitable for babies after six months include: puréed apricots, molasses, refined lentils, cereals, well mashed beans and green vegetables. avoid cereals that are very high in fibre as these may inhibit iron absorption. Vitamin C aids absorption of iron from plant foods and so it helps to give sources of these nutrients together. Vitamin C is found in frozen, fresh or juiced fruit and vegetables.

Calcium

Breast or formula milk contains all the calcium your baby needs initially. Good sources of calcium for the later weaning stages include cow's and fortified soya milk, cheese, green vegetables, wholemeal bread, beans, lentils, ground almonds, sesame paste and tofu.

Protein

Because babies are growing rapidly they require more protein than adults compared to their body weight. Breast or formula milk will provide the major source of protein for the first eight months. A variety of foods should be given each day so the baby obtains the right balance of amino acids. For example if a baby has toast, lentils and yoghurt during a day that would give a good balance.

Energy

Babies between the age of 6 and 12 months require 700 to 1000 calories a day, so they need concentrated sources of energy. Babies and young children do not have the capacity to eat large quantities of food and so they need small and frequent meals. Their diet should not contain too many foods that are bulky or watery. Make sure your baby has some concentrated energy foods like lentils with vegetable oil, avocado, cheese or smooth nut butter. Sugar is not a good source of energy for babies.

Vitamin B12

Vitamin B12 is made by micro-organisms and is found mostly in animal foods. Very young babies will get all the vitamin B12 they need from formula or breast milk. Later, vegetarian babies should obtain enough of this vitamin from dairy products and eggs. Vegan babies will need vitamin B12 from fortified foods such as some soya milks, low salt yeast extract or veggie burgers.

Vitamin D

Vitamin D is found in dairy products, eggs and fortified foods like margarine and some breakfast cereals, and can be made by the action of sunlight on the skin. It is found exclusively in animal foods so vegan babies may need a vitamin D supplement. Breast or formula milk should provide all the vitamin D needed initially.

Fibre

A diet too high in fibre will fill up a child before their nutritional needs have been met and can interfere with absorption of minerals, such as zinc, iron and calcium, so refined bran must not be added to a young child's diet. If you think your baby is constipated give extra fluid such as water or diluted fruit juices.

Milk

Babies under two should not be given semi-skimmed and children under five should not be given skimmed milk because skimmed milk lacks the fat soluble vitamins A and D and young children need the energy from fat. Soya milks should be specially formulated for babies if they are used instead of breast milk and should be fortified if used as an alternative to cow's milk for young children. It is recommended that if avoiding animal milk products, soya formula be used up to the age of 2 years.

Salt and Sugar

These should be avoided in the diet of babies and young children. A baby's kidneys are not mature enough to cope with too much salt, and sugary foods and drinks are a prime cause of tooth decay. Sugar gives calories without any associated vitamins or minerals. In addition, a baby who is encouraged to develop a sweet tooth may have problems with obesity in later life.

Nuts

Whole or chopped nuts and seeds are not suitable for children under five because of the danger of choking, but they can be used if finely ground, for example in cooking or smooth nut spread. However many experts suggest avoiding nut products altogether in a small child's diet due to the risk of allergies developing. If there is a history of allergies or other atopic diseases in a family it is certainly best to avoid any nuts, especially peanuts, until at least three years. The pregnant or breastfeeding mother should also avoid peanuts if there is a history of allergies in the family.

Quorn and Textured Vegetable Protein

Quorn products are a useful addition to the diet of young children, but it should not be relied on as the sole or major source of protein since it is relatively low in calories and high in fibre so may satisfy the child's appetite before they have taken in enough energy. Textured vegetable protein may be difficult for young babies to digest, and its salt content must be watched.

Further Information

Even with the help of this Info Sheet and the growing scientific evidence that a vegetarian diet is a healthy option, you may experience resistance from health professionals, family or friends about bringing up your baby as a vegetarian. Contact The Vegetarian Society for help in solving any problems or answering your questions.

PREGNANCY

Preconception

The health of a mother and baby is influenced not only by diet during pregnancy but also by diet before conception. Eating a healthy diet before pregnancy will give your body a good store of nutrients for the baby to draw on during pregnancy. The foetus is most susceptible to nutritional imbalance during the first few months of pregnancy because this is the time of most rapid development. If the lady is very underweight or overweight, she should try to achieve an acceptable weight for her height by a sensible and well-balanced diet.

If she uses to take the contraceptive pill, it is a good idea to come off and use an alternative method of contraception a few months before conceiving.

This is because the pill can alter levels of some nutrients, particularly vitamin B6, folate and zinc. Ensure that she have a well balanced diet with plenty of fresh fruit, vegetables (especially green vegetables) and wholegrain cereals. Try to avoid fatty foods, sweets, biscuits and cakes. Now is the time to make other healthy lifestyle changes and so stop smoking, cut down on alcohol and drinks containing caffeine such as tea, coffee and cola. Try a barley drink such as Barley Cup, or herb and fruit teas instead.

If the lady find that she is pregnant before she have had time to think about preconception, then don't worry, there is still plenty of time to make healthy changes to the diet.

Pregnancy is a time when good nutrition is vital, for vegetarians and non-vegetarians alike. It is a time of readjustment as well as growth. The nine months are divided up into three divisions of three months each called trimesters. Many women experience changes in mood, activity and appetite with the different stages of pregnancy.

There is no truth in the old saying that pregnancy means eating for two. The extra energy needed is only 200-300 calories a day for nine months. This is equivalent to two slices of bread with margarine or a jacket potato with baked beans or cheese.

Some women do feel a lot more hungry than this and if they are gaining weight at the right rate, they should eat according to their appetite.

0-3 Months

Women usually feel different and may experience tiredness as well as a feeling of sickness particularly in the morning. Calorie needs are only about 100kcal more in the early stages of pregnancy and some women do not experience much increase in appetite until the end of this period.

3-6 Months

Appetite usually increases after the first three months and this period requires about 300-400kcal extra calories a day. Extra calorie needs should be met by cereals, pulses, nuts and seeds, dairy products (unless vegan) and starchy vegetables such as potatoes.

6-9 Months

The baby is maturing now and this is a time for easing up on activity and preparing for the birth. The baby takes up a lot of space and may press on the stomach, reducing the capacity for food. Many women feel they need to eat small meals more frequently at this stage. Normal weight gain during pregnancy is about 22 to 28 pounds or one and a half to two stones (10-12.5kg). Weight gain often slows down during the last few months of pregnancy.

Exercise

It is a good idea to do some form of exercise during pregnancy. Antenatal clinics normally advise on appropriate exercise. Swimming or gentle yoga are often recommended.

Nutrients

Fluid

You may find that you are more thirsty during pregnancy. This is natural as fluid intake should increase. Never allow yourself to become over thirsty and include plenty of fresh water, dilute fruit juices, milk (soya or cow's) and herb teas. Drinks containing caffeine (tea, coffee and cola) should be limited and alcohol should be avoided altogether if possible.

Protein

Increased protein needs in pregnancy are usually met simply by the extra calories from more foods. Protein can be found in milk, cheese, eggs, soya milk, tofu, cereals, nuts and pulses. A normal variety of these foods will provide adequate protein. Intake of dairy products and eggs should not increase dramatically. Some people believe that excessive amounts may sensitise the baby in the womb to allergies towards these foods.

Iron

The need for iron is increased during pregnancy, especially during the later stages. Anaemia, due to iron deficiency, is common in pregnancy whether you are vegetarian or not. Vegetarians should be especially careful to include plenty of iron in their diet as vegetable sources are not as well absorbed.

Good vegetarian sources of iron can be found in wholegrain cereals, pulses, green vegetables and dried fruits. Iron absorption is increased if taken with a good source of vitamin C, which can be found in fresh fruit and vegetables.

Tea contains tannin which can inhibit iron absorption and should not be taken an hour before or after a meal.

Many doctors routinely prescribe iron tablets for pregnant women. Iron tablets may not be needed unless a blood test demonstrates anaemia. Iron levels normally decrease during pregnancy as the blood becomes more dilute. Some women prefer to take a natural iron supplement, such as Floridix, which is available from health food stores.

Calcium and Vitamin D

The body needs extra calcium during pregnancy, especially in the later stages, to enable the baby's bones to develop. Calcium absorption from the gut is more efficient during pregnancy and this should provide enough to meet requirements. Vegans and vegetarian women who consume few dairy products need to be particularly careful to ensure adequate calcium in the diet. Some vegan women, especially if they intend to breast-feed, may decide on a calcium supplement as a wise precaution, although with a good vegetable intake of calcium, it may not be necessary.

Good sources of calcium include green vegetables, almonds, sesame seeds or tahini, cow's milk, tofu, cheese, yoghurt, wholegrain cereals and pulses. Vitamin D is essential for calcium absorption and can be obtained from sunlight, margarine and dairy products.

Folate

Folate is one of the B vitamins needed in increased amounts during pregnancy. Research has shown that a deficiency of folate during pregnancy can lead to birth defects. Vegetarians should not be at risk as the best sources of this vitamin are green leafy vegetables, fruit, peanuts, yeast extract and wholegrain cereals.

Vitamin B12

Vitamin B12 is essential to the growth and development of your baby. If adequate amounts of dairy products, eggs and fortified yeast extract are included in your diet, then you should have enough vitamin B12. It is especially important for vegans to include a reliable source of vitamin B12 in the diet during pregnancy. Some vegan foods, such as certain brands of soya milk, margarine and soya products are fortified with this vitamin. If the vitamin B12 in your diet is unreliable, then a supplement is recommended.

WHAT FOODS AND HOW MUCH?

For example, a cereal serving can be two small slices of bread, a portion of pasta or a large potato. A fruit serving is equivalent to one whole piece and a vegetable serving is about 3.5oz (100g). A large portion of nutloaf could be counted as a cereal and a nut serving. If you are gaining weight normally, the size of a serving can be based on appetite. It is a good idea to avoid sweets, cakes, sugar and soft drinks as these provide extra calories without giving much in the way of nutrients.

Food Group	Servings / day	Provides
Cereals (wheat, oats, rice, barley, rye etc.)	Four	Energy, protein, fibre, calcium, B vitamins, iron, zinc
Pulses (beans, chick peas, lentils etc.)	One - two	Protein, fibre, energy
Nuts & seeds (almonds, peanuts, nutbutters, sesame seeds etc.)	One - two (small servings)	Energy, protein, fat, calcium
Fruit & vegetables; Leafy greens, Red & yellow vegetables, Fresh fruit, Dried fruit	Four - five	Iron, calcium, folate, Vitamin A, Vitamin C
Dairy products & Soya products (soya milk, tofu etc.)	Three - four	Protein, calcium, vitamin B12
Vegetable oils & fats	Two (small servings)	Energy, vitamins A, D (margarine & butter) & E (vegetable oils)
Yeast extract & other Vitamin B12 fortified foods	One	Vitamin B12

During pregnancy over-the-counter medicines and tablets should be avoided unless prescribed by your doctor. A vitamin and mineral supplement should not be needed if a good balanced vegetarian diet is followed, but will not cause any harm if taken as a precaution.

General Advice

Morning Sickness

Studies show that about three quarters of all women experience nausea and vomiting during pregnancy. Nausea normally occurs in the first few months although it can last throughout pregnancy.

Morning sickness can be relieved by having a dry biscuit or toast before getting up. Avoiding long intervals between meals helps, as nausea often occurs at the same time as hunger. Starchy foods, such as bread and potatoes, should be eaten regularly as they help maintain blood sugar level and fill the stomach, helping to relieve the sickness.

Food Safety

Pregnant women are advised to avoid soft cheeses such as Brie and mould ripened cheese such as Stilton, because of the risk of listeria. Cottage cheese or hard cheeses, such as Cheddar should be used instead. Even free range eggs have been found to contain Salmonella. Raw or lightly cooked eggs should be avoided.

Heartburn

Heartburn is very common in the later stages of pregnancy. It can be prevented by avoiding large meals and instead choosing small frequent meals or snacks. It can help to sit up very straight when eating and avoid activity just after a meal. Spicy and fatty foods, fizzy drinks and citrus fruits can make the problem worse.

Constipation

During pregnancy, the digestive system absorbs nutrients more efficiently and this can contribute to constipation. Iron tablets often make the problem worse. Constipation can be relieved by increasing fluid intake and including plenty of wholegrain cereals, pulses, fruits and vegetables in the diet.

Food Cravings and Aversions

Many women experience altered taste preferences during pregnancy which vary considerably. Some experience very strong cravings for particular foods such as fruit, salad vegetables, nuts, starchy foods or chocolate. Aversions to fatty foods, alcohol, tea and coffee are also common. A healthy vegetarian diet can provide you with all the nutrients you need during your pregnancy and give your baby the best possible start in life.

Iron

Iron is an essential component of haemoglobin, transporting oxygen in the blood to all parts of the body. It also plays a vital role in many metabolic reactions. Iron deficiency can cause anaemia resulting from low levels of haemoglobin in the blood. Iron deficiency is the most widespread mineral nutritional deficiency both in Britain and worldwide.

Functions

Iron is essential for the formation of haemoglobin, the red pigment in blood. The iron in haemoglobin combines with oxygen and transports it through the blood to the body's tissues and organs. The body contains between 3.5 and 4.5g of iron, 2/3 of which is present in haemoglobin. The remainder is stored in the liver, spleen and bone-marrow. A small amount is present as myoglobin, which acts as an oxygen store in muscle tissue.

Iron deficiency can lead to anaemia. Iron stores in the body become depleted and haemoglobin synthesis is inhibited. Symptoms of anaemia include tiredness, lack of stamina, breathlessness, headaches, insomnia, loss of appetite and pallor.

All these symptoms are associated with decreased oxygen supply to tissues and organs. Iron also plays an important role in the immune system, people with low iron levels having lowered resistance to infection.

Research has also shown iron deficiency to be associated with impaired brain function, and iron deficiency in infants can result in impaired learning ability and behavioural problems. Iron deficiency is the most prevalent nutritional problem both in Britain and worldwide.

It has been stated that 2/3 of children and women of child-bearing age in developing countries suffer from iron deficiency, 1/3 suffering from severe deficiency and anaemia. In developed countries, between 10-20% of child-bearing age women are said to be anaemic. Iron is the least plentiful nutrient in the typical British diet and anaemia is fairly common in the UK.

Dietary Sources

Dietary iron exists in two different forms. Haem iron only exists in animal tissues, whilst in plant foods iron is present as non-haem iron. In a mixed omnivore diet around 25% of dietary iron is non-haem iron. Non-haem iron is less easily absorbed by the body than is haem iron. The amount of iron absorbed from various foods ranges from around 1 to 10% from plant foods and 10 to 20% from animal foods.

The absorption of iron is influenced by other constituents of a meal. Phytates, oxalates and phosphates present in plant foods can inhibit absorption, as can tannin in tea. Fibre may also inhibit absorption. Vitamin C greatly increases the absorption of non-haem iron. Foods rich in vitamin C include citrus fruits, green peppers, and fresh leafy green vegetables. Citric acid, sugars, amino acids and alcohol can also promote iron absorption.

Iron absorption can also be influenced by the amount of iron in the diet. Lowered levels of iron in the diet result in improved absorption. Good sources of iron for vegetarians include wholegrain cereals and flours, leafy green vegetables, blackstrap molasses, pulses such as lentils and kidney beans, and some dried fruits.

Sources of Iron (single servings)					
Good sources		**Fair sources**		**Poor sources**	
Chick peas (200g or 7oz)	6.2 mg	Egg, boiled	1.3 mg	Banana (120g or 4¼oz)	0.48 mg
Bran flakes (45g or 1½oz)	5.3 mg	Avocado (75g or 2¾oz)	1.1 mg	Yoghurt (150g or 5½oz)	0.36 mg
Spinach, boiled (100g or 3½oz)	4.0 mg	Asparagus (125g or 4¾oz)	1.1 mg	Cow's milk (½ pint)	0.14 mg
Baked beans (225g or 8oz)	3.2 mg	1 slice wholemeal bread (40g)	1.0 mg	Hard cheese (30g or 1oz)	0.12 mg
Black treacle (35g or 1¼oz)	3.2 mg	Broccoli, boiled (100g or 3½oz)	1.0 mg	Margarine (7g or ¼oz)	0.02 mg
Muesli (60g or 2¼oz)	2.76mg	Brown rice (200g or 7oz)	0.9 mg	.	.
4 Dried figs (60g or 2oz)	2.1 mg	Peanut butter (20g or ½oz)	0.5 mg	.	.
8 Dried apricots (50g or 1¾oz)	2.1 mg	.	.	.	.

Despite iron from plant foods being less readily absorbed research has shown that vegetarians are no more likely to suffer from iron deficiency than non-vegetarians. Draper & Wheeler (1989) have stated there is no indication of increased prevalence of iron deficiency amongst vegetarians. Anderson (1981) found the iron status of long-term vegetarian women to be adequate, despite a high intake of fibre and phytate.

Required Intakes

The old Recommended Daily Amounts (RDA's) have now been replaced by the term Reference Nutrient Intake (RNI). The

RNI is the amount of nutrient which is enough for at least 97% of the population.

Nutrient Intakes for Iron, mg/day

Age	RNI	Age	RNI
0 to 3 months	1.7 mg	Men 11-18 yrs	11.3 mg
4 to 6 months	4.3 mg	Men 19 + yrs	8.7 mg
7 to 12 months	7.8 mg	Women 11-49 yrs	14.8 mg
1 to 3 yrs	6.9 mg	Women 50 + yrs	8.7 mg
4 to 6 yrs	6.1 mg	-	-
7 to 10 yrs	8.7 mg	-	-

In women of child-bearing age, loss of iron from menstruation of blood adds considerably to iron need. These losses can be highly variable. Around 10% of women of child-bearing age will need more iron than is indicated. In men, post-menopausal women, and children iron is efficiently conserved by the body. Iron in haemoglobin is recycled and the amount of iron lost from the body is very small. Infants and children need extra iron to increase blood volume and muscle tissue. Extra iron is also required during pregnancy and breast feeding.

Meal Plan

Sample one day's meal plan to meet the RNI of 14.8 mg for a women aged 11 to 49 years					
Breakfast		**Lunch**		**Evening meal**	
Bowl of muesli with milk	2.76mg	2 slices of wholemeal bread	2.0 mg	Brown rice (200g or 7oz)	0.94mg
1 slice of toast	1.0 mg	Peanut butter (20g or ¾oz)	0.5 mg	Chick peas (200g or 7oz)	6.2 mg
-	-	1 banana	0.5 mg	Broccoli (100g or 3½oz)	1.0 mg
Total iron intake - 14.9 mg					

OMEGA 3 FATS

So how does current government advice that "we" should try to eat at least two servings of fish a week, including one serving of oily fish, as part of a healthy balanced diet translate into dietary advice for vegetarians? And what should you do if your GP suggests fish oils for creaking joints or to thin your blood? Well, the first thing to remember is that the vegetarian diet is widely recognised as being protective against heart disease, the main ailment that the advice is aimed at, so vegetarians have a head start already, and of course the general population don't exactly eat much oily fish, or indeed offal, the other "good" source of omega 3 fats.

To begin with we must distinguish between the two polyunsaturated fatty acids which are termed essential because they can not be made in the body and therefore must be present in the diet. They are linoleic acid (LA), an omega 6 fat, which is widely available in a vegetarian diet, being present in large quantities in most oils and other vegetable based fatty foods, and alpha-linolenic acid (ALA), an omega 3 fat, which is not so widely available in a vegetarian diet, and is generally considered to be the more beneficial of the two EFAs.

Alpha-linolenic acid is what is known as an omega 3 fat, and is a precursor of the longer chain omega 3 fats eicosapentaenoic acid (EPA) and docosahexaenoic acid (DHA)-i.e., EPA and to a lesser extent DHA can be made in the body from ALA. These two fatty acids are the ones available in significant amounts in oily fish, and fish oil supplements. All three omega 3 fats have been shown to offer numerous physiological benefits, notably their anti-inflamatory properties and their ability to offer cardioprotective effects especially in people with pre-existing cardiovascular problems, though EPA and DHA are more potent than simple ALA.

Generally, vegetarian, and especially vegan, diets are relatively low in ALA compared with LA, and provide little EPA and DHA directly (though a certain amount of DHA is found in eggs, especially from hens fed on flax seeds or algae), and tissue levels of long chain omega 3 fatty acids have been shown to be relatively low in vegetarians and vegans, even though ALA intake varies little across vegans, vegetarians and omnivores.

Taking an overview of the various fatty acids intake recommendations worldwide, and the confounding factors surrounding the common vegetarian diet, leads to a conclusion that an ALA intake of 1.5% of total energy is optimum for vegetarians-or roughly 4g a day. This should provide enough of the parent omega 3 fat to ensure significant amounts of EPA and DHA are formed by the body (conversion rates are around 5-10% for EPA and 2-5% for DHA). However it is also important for vegetarians to ensure that their intake of LA is not too high compared with ALA since a higher intake of LA interferes with the process in which the human body converts ALA into the even more beneficial EPA and DHA, so a LA to ALA ratio of around 4 to 1 or slightly lower is considered to be the optimum, but any steps to bring down an excessively high amount of omega 6 fats in the diet would be beneficial.

Sources of Omega 3 Fats		
Flaxseed oil	1 tablespoon (14g)	provides 8.0g of ALA and relatively insignificant levels of LA
Flaxseed, ground	1 tablespoon (24g)	provides 3.8g of ALA
Rape Seed oil	1 tablespoon (14g)	provides 1.6g of ALA and only twice as much LA
Walnuts	1 oz (28g)	provides 2.6g of ALA but also four times as much LA
Tofu	4.5oz (126g)	provides 0.7g of ALA but also seven times as much LA

Practical Steps

There are a number of steps to take to ensure that the optimum levels of all the omega 3 fats are present in the body.

1. Make sure you include a good source of ALA in your diet, the simplest source would be one teaspoon of flax seed oil a day, taken either on its own or mixed into dressings etc. Flax oil is also available in vegetable capsules. Alternately include 4 to 5 teaspoons of ground flax seeds, or rape seed oil in your diet-though do not heat any of the oils, and only add the flax seeds to any foods at a late stage since heating will destabilise the ALA. It is important that the flax seeds are ground or at least crushed, if left whole much of the fat will be unavailable.

2. Replace fats high in omega 6 oils, such as sunflower oil or corn oil, with fats higher in monounsaturated fats, such as olive oil or rape seed oil which do not disrupt the formation of EPA and DHA.
3. Other foods can add to your intake of ALA. Most of the little fat in leafy green vegetables is ALA-broccoli has 0.13g per 100g, cabbage 0.11g per 100g, so simply eating your greens is making a positive addition to your intake. Walnuts and tofu are also good sources but are comparably high in LA.

Pregnant or nursing mothers who are uncertain whether their diet is providing enough omega 3 fats may wish to consider supplementing their diet with a direct source of DHA since this appears to play an important part in the development of immature brains. DHA supplements derived from algae and encased in non gelatine capsules are now available. It has also been suggested that DHA supplements may be of help to children with certain behavioral or learning difficulties.

Protein

Being vegetarian does not mean your diet will be lacking in protein. Most plant foods contain protein and in fact it would be very difficult to design a vegetarian diet that is short on protein. Excess dietary protein may lead to health problems. It is now thought that one of the benefits of a vegetarian diet is that it contains adequate but not excessive protein. Proteins are made up of smaller units called amino acids. There are about 20 different amino acids, eight of which must be present in the diet. These are the essential amino acids. Unlike animal proteins, plant proteins may not contain all the essential amino acids in the necessary proportions. However, a varied vegetarian diet means a mixture of proteins are consumed, the amino acids in one protein compensating for the deficiencies of another.

Structure and Functions

Proteins are highly complex molecules comprised of linked amino acids. Amino acids are simple compounds containing carbon, hydrogen, oxygen, nitrogen and occasionally sulphur. There are about 20 different amino acids commonly found in plant and animal proteins. Amino acids link together to form chains called peptides. A typical protein may contain 500 or more amino acids. Each protein has it's own unique number and sequence of amino acids which determines it's particular structure and function. Proteins are broken down into their constituent amino acids during digestion which are then absorbed and used to make new proteins in the body. Certain amino acids can be made by the human body. However, the essential amino acids cannot be made and so they must be supplied in the diet. The eight essential amino acids required by humans are: leucine, isoleucine, valine, threonine, methionine, phenylalanine, tryptophan, and lysine. For children, histidine is also considered to be an essential amino acid.

Proteins are essential for growth and repair. They play a crucial role in virtually all biological processes in the body. All enzymes are proteins and are vital for the body's metabolism. Muscle contraction, immune protection, and the transmission of nerve impulses are all dependent on proteins. Proteins in skin and bone provide structural support. Many hormones are proteins. Protein can also provide a source of energy. Generally the body uses carbohydrate and fat for energy but when there is excess dietary protein or inadequate dietary fat and carbohydrate, protein is used. Excess protein may also be converted to fat and stored.

Dietary Sources

Most foods contain at least some protein. Good sources of protein for vegetarians include nuts and seeds, pulses, soya products (tofu, soya milk and textured soya protein such as soya mince), cereals (wheat, oats, and rice), free-range eggs and some dairy products (milk, cheese and yoghurt).

Different foods contain different proteins, each with their own unique amino acid composition. The proportions of essential amino acids in foods may differ from the proportions needed by the body to make proteins. The proportion of each of the essential amino acids in foods containing protein determines the quality of that protein. Dietary proteins with all the essential amino acids in the proportions required by the body are said to be a high quality protein. If the protein is low in one or more of the essential amino acids the protein is of a lower quality. The amino acid that is in shortest supply is called the limiting amino acid.

Sources of protein (single servings)					
Good sources		**Fair Sources**		**Poor sources**	
Chick peas (200g or 7oz)	16.0g	Brown rice (200g or 7oz)	4.4g	1 Carrot	0.4g
Baked beans (225g or 8oz)	11.5g	Broccoli (100g or 3½oz)	3.1g	1 Apple	0.3g
Tofu (140g or 5oz)	10.3g	Potatoes (200g or 7oz)	2.8g	Cream, double (20g or 2/3oz)	0.3g
Cow's milk (½ pint)	9.2g	Porridge [water] (160g or 6oz)	2.4g	Butter/margarine	None
Lentils (120g or 4¼oz)	9.1g	-	-	Vegetable oil	None
Soya milk (½ pint)	8.2g	-	-	Sugar or syrup	None
Muesli (60g or 2¼oz)	7.7g	-	-	-	-
Egg, boiled	7.5g	-	-	-	-
Peanuts (30g or 1oz)	7.3g	-	-	-	-
Bread, (2 slices)	7.0g	-	-	-	-
Hard cheese (30g or 1oz)	6.8g	-	-	-	-

Protein quality is usually defined according to the amino acid pattern of egg protein, which is regarded as the ideal. As such, it is not surprising that animal proteins, such as meat, milk and cheese tend to be of a higher protein quality than plant proteins. This is why plant proteins are sometimes referred to as low quality proteins. Many plant proteins are low in one of the essential amino acids.

For instance, grains tend to be short of lysine whilst pulses are short of methionine. This does not mean that vegetarians or vegans go short on essential amino acids. Combining plant proteins, such as a grain with a pulse, leads to a high quality protein which is just as good, and in some cases better, than protein from animal foods. Soya is a high quality protein on its own which can be regarded as equal to meat protein.

The limiting amino acid tends to be different in different proteins. This means when two different foods are combined, the amino acids in one protein can compensate for the one lacking in the other. This is known as protein complementing. Vegetarians and vegans eating a well-balanced diet based on grains, pulses, seeds, nuts and vegetables will be consuming a mixture of proteins that complement one another naturally without requiring any planning. Beans on toast, cheese or peanut butter sandwich, muesli with milk (soya or cow's), and rice with peas or beans are all common examples of protein complementing. Previously, it has been thought that protein complementing needed to occur within a single meal. However, it is now known that this is not necessary as the body keeps a short-term store of the essential amino acids. A well-balanced vegetarian or vegan will easily supply all the protein and essential amino acids needed by the body.

Vegan Nutrition

A vegan is a strict vegetarian who does not eat any dairy products or eggs. Most vegans do not eat honey. A well balanced vegan diet can provide all the essential nutrients you require and shares the same health advantages as a vegetarian diet. Nutritional guidelines for vegans are essentially similar to those for vegetarians. However, vegetarians gain certain nutrients from dairy products and eggs. Vegans need to ensure their diets contain plant food sources of these nutrients.

Protein

Obtaining adequate protein on a vegan diet is not a problem. Nuts & seeds, pulses, wholegrain and grain products and soya products all supply protein. Previously, it has been thought that plant proteins are of a lower quality than animal proteins in terms of their essential amino acid content. However, this is no longer regarded as a problem and eating a balanced diet of plant foods will provide all the essential amino acids in adequate amounts.

Essential Fatty Acids

There are two essential fatty acids which must be supplied by the diet. These are linoleic acid and a-linolenic acid. Essential fatty acids are important for cell membrane function, cholesterol metabolism and the synthesis of various metabolites. Good sources of essential fatty acids are vegetable oils. It is important to have the correct balance between linoleic acid and a-linolenic acid. It has been suggested that vegans should use soyabean or rapeseed oils rather than sunflower or corn oils as these help give a better dietary balance.

Vitamin B2 (Riboflavin): Certain studies have found vegans to have a low intake of the vitamin, riboflavin. Riboflavin is important in converting protein, fats and carbohydrates into energy, and the synthesis and repair of body tissues. Good sources of riboflavin include whole grains, mushrooms, almonds, leafy green vegetables and yeast extracts.

Vitamin B12: Vitamin B12 is found primarily in meat, dairy products and eggs and is absent from plant foods. Considerable research has been carried out into possible plant sources of B12. Fermented soya products, seaweeds and algae such as spirulina have all been proposed as containing significant amounts of B12. However, the present consensus is that any B12 present in plant foods is likely to be in a form unavailable to humans and so these foods should not be relied upon as safe sources.

Vitamin B12 is important in the formation of red blood cells and the maintenance of a healthy nervous system. When deficiency does occur it is more likely to be due to a failure to absorb B12 from the intestine than a dietary deficiency.

Vegans can obtain B12 from a wide range of foods which have been fortified with the vitamin. These include certain yeast extracts, veggieburger mixes, breakfast cereals, vegetable margarines and soya milks. You should check the packaging to see which individual products are fortified with B12.

Vitamin D: Vitamin D is present in oily fish, eggs and dairy products in variable amounts. It is not found in plant foods. However, vegans can obtain vitamin D from vegetable margarines, some soya milks and certain other foods which are fortified with the vitamin.

Vitamin D is also synthesised by the skin when exposed to sunlight. Synthesis of vitamin D in this way is usually adequate to supply all the body's requirements. Most vegans will obtain sufficient vitamin D providing they spend time outdoors on bright days. Fortified foods further ensure adequate amounts.

Vegans who may be confined indoors may be recommended a vitamin D supplement. Also, infants who are seldom outdoors or who are dark-skinned may require supplements. Asian vegans may also be at risk of deficiency, particularly Asian women who may be required to keep their skin covered for cultural reasons.

Calcium: The major source of calcium in British diets is generally milk and dairy products. Vegans can obtain adequate calcium from plant foods. Good sources include tofu, leafy green vegetables, watercress, dried fruit, seeds and nuts. Also, white bread is fortified with calcium, as are some soya milks. Hard water can also provide significant amounts of calcium.

Iodine: Milk is the primary source of iodine in the British diet and studies have indicated some vegans may have a low iodine intake. Seaweeds are a good source of iodine, and vegetables and grains can contain iodine depending on the amounts in the soil.

Infants: It is perfectly possible to bring up a child on a vegan diet. Vegan children should be given plenty of nutrient rich foods and need good sources of protein, calcium, vitamin B12 and vitamin D. High fibre foods can fill up a child without filling their nutritional needs as well as interfering with mineral absorption from the intestine. For these reasons, foods high in fibre shouldn't be overused.

Vegan Storecupboard: Dairy products can largely be replaced with various soya products. There are several brands of soya milk. It can be purchased either sweetened or unsweetened, plain or flavoured. Different brands may be fortified with vitamin B12, vitamin D and calcium. Soya cheeses, yoghurts and cream are all available from health food stores. Eggs can be replaced in recipes by commercial egg replacer products, also available from health food stores.

Vitamin B12

Vitamin B12 is a member of the vitamin B complex. It contains cobalt, and so is also known as cobalamin. It is exclusively synthesised by bacteria and is found primarily in meat, eggs and dairy products.

There has been considerable research into proposed plant sources of vitamin B12. Fermented soya products, seaweeds, and algae such as spirulina have all been suggested as containing significant B12.

However, the present consensus is that any B12 present in plant foods is likely to be unavailable to humans and so these foods should not be relied upon as safe sources. Many vegan foods are supplemented with B12.

Vitamin B12 is necessary for the synthesis of red blood cells, the maintenance of the nervous system, and growth and development in children. Deficiency can cause anaemia. Vitamin B12 neuropathy, involving the degeneration of nerve fibres and irreversible neurological damage, can also occur.

Functions

Vitamin B12's primary functions are in the formation of red blood cells and the maintenance of a healthy nervous system. B12 is necessary for the rapid synthesis of DNA during cell division. This is especially important in tissues where cells are dividing rapidly, particularly the bone marrow tissues responsible for red blood cell formation.

If B12 deficiency occurs, DNA production is disrupted and abnormal cells called megaloblasts occur. This results in anaemia. Symptoms include excessive tiredness, breathlessness, listlessness, pallor, and poor resistance to infection. Other symptoms can include a smooth, sore tongue and menstrual disorders. Anaemia may also be due to folic acid deficiency, folic acid also being necessary for DNA synthesis.

B12 is also important in maintaining the nervous system. Nerves are surrounded by an insulating fatty sheath comprised of a complex protein called myelin. B12 plays a vital role in the metabolism of fatty acids essential for the maintenance of myelin. Prolonged B12 deficiency can lead to nerve degeneration and irreversible neurological damage.

When deficiency occurs, it is more commonly linked to a failure to effectively absorb B12 from the intestine rather than a dietary deficiency. Absorption of B12 requires the secretion from the cells lining the stomach of a glycoprotein, known as intrinsic factor. The B12-intrinsic factor complex is then absorbed in the ileum (part of the small intestine) in the presence of calcium. Certain people are unable to produce intrinsic factor and the subsequent pernicious anaemia is treated with injections of B12.

Vitamin B12 can be stored in small amounts by the body. Total body store is 2-5mg in adults. Around 80% of this is stored in the liver.

Vitamin B12 is excreted in the bile and is effectively reabsorbed. This is known as enterohepatic circulation. The amount of B12 excreted in the bile can vary from 1 to 10ug (micrograms) a day.

People on diets low in B12, including vegans and some vegetarians, may be obtaining more B12 from reabsorption than from dietary sources. Reabsorption is the reason it can take over 20 years for deficiency disease to develop in people changing to diets absent in B12. In comparison, if B12 deficiency is due to a failure in absorption it can take only 3 years for deficiency disease to occur.

Dietary Sources

The only reliable unfortified sources of vitamin B12 are meat, dairy products and eggs. There has been considerable research into possible plant food sources of B12. Fermented soya products, seaweeds and algae have all been proposed as possible sources of B12. However, analysis of fermented soya products, including tempeh, miso, shoyu and tamari, found no significant B12. Spirulina, an algae available as a dietary supplement in tablet form, and nori, a seaweed, have both appeared to contain significant amounts of B12 after analysis. However, it is thought that this is due to the presence of compounds structurally similar to B12, known as B12 analogues. These cannot be utilised to satisfy dietary needs. Assay methods used to detect B12 are unable to differentiate between B12 and it's analogues, Analysis of possible B12 sources may give false positive results due to the presence of these analogues.

Researchers have suggested that supposed B12 supplements such as spirulina may in fact increase the risk of B12 deficiency disease, as the B12 analogues can compete with B12 and inhibit metabolism. The current nutritional consensus is that no plant foods can be relied on as a safe source of vitamin B12.

Bacteria present in the large intestine are able to synthesise B12. In the past, it has been thought that the B12 produced by these colonic bacteria could be absorbed and utilised by humans. However, the bacteria produce B12 too far down the intestine for absorption to occur, B12 not being absorbed through the colon lining.

Human faeces can contain significant B12. A study has shown that a group of Iranian vegans obtained adequate B12 from unwashed vegetables which had been fertilised with human manure. Faecal contamination of vegetables and other plant foods can make a significant contribution to dietary needs, particularly in areas where hygiene standards may be low. This may be responsible for the lack of aneamia due to B12 deficiency in vegan communities in developing countries.

Good sources of vitamin B12 for vegetarians are dairy products or free-range eggs. ½ pint of milk (full fat or semi skimmed) contains 1.2 μg. A slice of vegetarian cheddar cheese (40g) contains 0.5 μg. A boiled egg contains 0.7 μg. Fermentation in the manufacture of yoghurt destroys much of the B12 present. Boiling milk can also destroy much of the B12. Vegans are recommended to ensure their diet includes foods fortified with vitamin B12. A range of B12 fortified foods are available. These include yeast extracts, Vecon vegetable stock, veggieburger mixes, textured vegetable protein, soya milks, vegetable and sunflower margarines, and breakfast cereals.

Required Intakes

The old Recommended Daily Amounts (RDA's) have now been replaced by the term Reference Nutrient intake (RNI). The RNI is the amount of nutrient which is enough for at least 97% of the population.

Reference Nutrient Intakes for Vitamin B12, μg/day. (1000 μg = 1mg)

Age	RNI
0 to 6 months	0.3 μg
7 to 12 months	0.4 μg
1 to 3 yrs	0.5 μg
4 to 6 yrs	0.8 μg
7 to 10 yrs	1.0 μg
11 to 14 yrs	1.2 μg
15 + yrs	1.5 μg
Breast feeding women	2.0 μg

Pregnant women are not thought to require any extra B12, though little is known about this. Lactating women need extra B12 to ensure an adequate supply in breast milk.

B12 has very low toxicity and high intakes are not thought to be dangerous.

6

VITAMINS, MINERALS AND OTHER NUTRIENTS

Vitamins

Vitamins are organic substances that are required in small amounts in the diet. They are necessary for numerous special functions in the body and are essential for good health. They can be affected by environmental conditions such as light, heat and air. Food storage, processing and cooking can all act to reduce the level of vitamins in food.

Vegetarians can obtain all the vitamins they need from a balanced diet. Vitamins may be either fat-soluble or water-soluble. The fat-soluble vitamins are vitamins A, D, E and K. Fat-soluble vitamins can be stored in the body and so dietary sources are not needed every day.

The water-soluble vitamins are vitamin C and the B group of vitamins. The B group of vitamins includes B1, B2, B3, B6, B12, folic acid, biotin and pantothenic acid.

The body is less able to store water-soluble vitamins (with the exception of vitamin B12 which is stored in the liver) and so they are needed daily. Water-soluble vitamins are more likely to be lost during cooking.

Vitamin A (retinol): For healthy skin, growth of bones, resistance to infection and night vision. Found in carrots, spinach, peppers, butter, margarine, watercress, dried apricots, full-fat dairy products. In plant foods it is present as its precursor, beta-carotene.

Vitamin B1 (thiamin): For breaking down carbohydrates for energy. Found in yeast extract, brazil nuts, peanuts, rice, bran, oatmeal, flour, wholemeal bread, sunflower seeds.

Vitamin B2 (riboflavin): Helps convert proteins, fats and carbohydrates into energy and for the growth and repair of tissues and healthy skin. Found in almonds, cheese, wholemeal bread, dried prunes, mushrooms, cashews, millet, avocados.

Vitamin B3 (niacin): For energy production, healthy skin, and the nervous system. Found in most foods including yeast extract, peanuts, wholemeal bread, mushrooms, sesame seeds.

Vitamin B6 (pyridoxine): For red blood cell formation and protein metabolism. Found in bran, wholemeal flour, yeast extract, hazelnuts, bananas, peanuts, currants.

Vitamin B12: For red blood cell formation, growth, and a healthy nervous system. Found in eggs, dairy products and fortified plant foods including soya milks, breakfast cereals, veggieburger mixes, yeast extracts and herbal soft drinks.

Folic Acid or Folate: For red blood cell formation, protein synthesis and DNA metabolism. Some functions are linked with vitamin B12. Found in yeast extract, spinach, broccoli, peanuts, almonds, hazelnuts.

Biotin: For energy production and healthy skin. Found in yeast extract, pulses, nuts, most vegetables.

Pantothenic Acid: For energy production and antibody formation. Widely found in most foods.

Vitamin C (Ascorbic acid): For healthy skin, bones, teeth and gums, resistance to infection and wound healing, energy production and growth. Found in citrus fruits, broccoli, spinach, berries, peppers.

Vitamin D: For the absorption of calcium and phosphate and healthy bones and teeth. Found in dairy products and margarine. Also produced by the action of sunlight on the skin. The elderly, young children and anyone confined indoors may need to consider a vitamin D supplement especially if vegan.

Vitamin E: Acts as an antioxidant protecting vitamins A and C and other important substances in the body. Found in vegetable oils, wheatgerm, hazelnuts, avocados.

Vitamin K: For effective blood clotting. Found in spinach, cabbage, cauliflower. Vitamin K is also obtained from bacterial synthesis in the intestine. Deficiency is rare.

Minerals

The main minerals required in the diet are calcium, magnesium, sodium, potassium, chlorine, phosphorus and sulphur. Other minerals which are required in only tiny quantities (less than 100mg/day) are called trace elements.

Sodium and Potassium: Both are important in maintaining the body's water balance controlling the composition of blood and other body fluids. Sodium chloride (salt) is present in processed foods and in small amounts in vegetables, fruits and grains. Most people consume too much sodium which can lead to high blood pressure. Potassium is widely found in plant foods, especially root vegetables and wholegrain cereals.

Calcium: For building and maintaining strong bones and teeth, muscle contraction and blood clotting. Found in dairy products, leafy green vegetables, almonds, sesame seeds, dried fruit, pulses, fortified soya milks.

Magnesium: For strong bones and enzymes involved in energy production. Widely found in plant foods and deficiency is very rare.

Phosphorus: Required with calcium for strong bones and teeth, muscle function and a vital component of all body cells. Found in nearly all foods and dietary deficiency is unknown.

Sulphur: Plays a role in some enzyme systems. Most dietary sulphur is in the sulphur-containing amino acids. There is no indication of sulphur deficiency except in association with protein deficiency.

Iron: An essential component of haemoglobin which transports oxygen in the blood through the body. Iron deficiency is one of the most common nutritional problems in a typical British diet. Research shows vegetarians are no more prone to iron deficiency than meat-eaters. Found in leafy green vegetables, pulses, wholemeal bread, dried fruit, pumpkin seeds, molasses. Vitamin C helps to absorb iron. Tea can reduce the amount of iron absorbed.

Zinc: Plays a role in a wide range of enzyme systems and is essential for DNA metabolism and growth. Vegetarian diets tend to contain less zinc than meat-based diets. Found in sesame and pumpkin seeds, green vegetables, cheese, lentils, wholegrain cereals.

Copper: For red blood cell formation and many enzyme functions. Found in green vegetables, yeast, nuts, wheatgerm.

Iodine: For the production of thyroid hormones important in body metabolism. Found in milk (used as a disinfectant in milk production), seafoods. Amount in plant foods depends upon the amount in soil in which plants were grown.

Selenium: For red blood cell function and acts as an antioxidant. Selenium in plant foods depends on selenium in the soil and can vary considerably.

Chromium: Necessary to maintain blood glucose level. Exact nutritional role is uncertain. Good sources include black pepper, yeast and wholemeal bread.

Fluorine: For strong bones and teeth. Main dietary source is fluoridated water also tea.

Manganese: For the function of many enzymes and also muscle function. Deficiency has never been observed. Found in tea, green vegetables, wholegrain cereals, nuts, spices.

Molybdenum: Essential part of some enzyme systems though deficiency has never been observed. Amount in plant foods depends on the amount in the soil.

Other Elements: Vanadium, boron, nickel, silicon and other elements are present in very small amounts in the body. There is some debate on whether they have any nutritional role. All the elements can be toxic if taken in excess. Some elements such as aluminium, cadmium, mercury and arsenic have no known nutritional role and are toxic at very low levels. These toxic Elements are often environmental pollutants.

ZINC

The mineral zinc is present in every part of the body and has a wide range of functions. It helps with the healing of wounds and is a vital component of many enzyme reactions. Zinc is vital for the healthy working of many of the body's systems. It is particularly important for healthy skin and is essential for a healthy immune system and resistance to infection.

Functions

Zinc has a range of functions. It plays a crucial role in growth and cell division where it is required for protein and DNA synthesis, in insulin activity, in the metabolism of the ovaries and testes, and in liver function. As a component of many enzymes, zinc is involved in the metabolism of proteins, carbohydrates, lipids and energy.

Our body contains about 2-3g of zinc. There are no specific storage sites known for zinc and so a regular supply in the diet is required. Zinc is found in all parts of our body, 60% is found in muscle, 30% in bone and about 5% in our skin. Particularly high concentrations are in the prostate gland and semen.

Men need more zinc than women because male semen contains 100 times more zinc than is found in the blood. The more sexually active a man the more zinc he will require. The recommended amounts of zinc for adult men are 1/3 higher than those for women.

The first signs of zinc deficiency are impairment of taste, a poor immune response and skin problems. Other symptoms of zinc deficiency can include hair loss, diarrhoea, fatigue, delayed wound healing, and decreased growth rate and mental development in infants.

It is thought that zinc supplementation can help skin conditions such as acne and eczema, prostate problems, anorexia nervosa, alcoholics and those suffering from trauma or post-surgery. It is always better to seek the advice of an expert before dosing yourself with supplements. If you choose to take a zinc supplement you should not need more than the daily recommended amount unless medical advice says otherwise.

Dietary Sources

Zinc is present in a wide variety of foods, particularly in association with protein foods. A vegetarian diet often contains less zinc than a meat based diet and so it is important for vegetarians to eat plenty of foods that are rich in this vital mineral.

Good sources for vegetarians include dairy products, beans and lentils, yeast, nuts, seeds and wholegrain cereals. Pumpkin seeds provide one of the most concentrated vegetarian food sources of zinc.

Only 20% of the zinc present in the diet is actually absorbed by the body. Dietary fibre and phytic acid, found in bran, wholegrain cereals, pulses and nuts, inhibit zinc absorption. Phytic acid forms a highly insoluble complex with zinc which the body cannot absorb. Cooking processes can reduce the adverse effects of both phytic acid and dietary fibre on zinc absorption.

Sources of zinc (single servings)

Good sources		Fair sources		Poor sources	
Chickpeas (200g or 7oz)	2.8mg	Peanut butter (20g or 2/3oz)	0.6mg	Dried dates (15g or 1/2oz)	0.04mg
Baked beans (225g or 8oz)	1.6mg	Peas, frozen/canned (80g or 2.5oz)	0.6mg	Butter (7g or 1/4oz)	0.01mg
1 Vegeburger (100g or 3.5oz)	1.6mg	3 Dried figs (60g or 2oz)	0.5mg	Raisins & Sultanas (15g or 1/2oz)	0.01mg
Pumpkin seeds (20g or 2/3oz)	1.3mg	3 Brazil nuts (10g or 1/3oz)	0.4mg		
Muesli (60g or 2.25oz)	1.3mg	Potatoes, boiled (200g or 7oz)	0.4mg		
Cheddar cheese (30g or 1oz)	1.2mg	1 Orange (140g or 5oz)	0.3mg		
Tahini paste (20g or 2/3oz)	1.1mg	6 Almonds (10g or 1/3oz)	0.3mg		
1 Fruit yoghurt (150g or 5.25oz)	0.9mg				
Dried yeast (10g or 1/3oz)	0.8mg				
Wholemeal bread (40g or 1.5oz)	0.8mg				

Baking can destroy over half the phytic acid in wholemeal bread. Zinc deficiency occurs where a large part of the diet consists of unleavened bread, such as Iran and other Middle East countries. High levels of the toxic mineral cadmium can also prevent zinc absorption because these two minerals compete for absorption. Conversely high levels of zinc in the diet can prevent the absorption of cadmium. Various chemicals added to many processed foods can also reduce zinc absorption eg phosphates, EDTA. A deficiency of zinc in the diet means zinc absorption is improved. Zinc is lost via the faeces, urine, hair, skin, sweat, semen and also menstruation.

Required Intakes

The old Recommended Daily Amounts (RDAs) have now been replaced by the term Reference Nutrient Intake (RNI). The RNI is the amount of nutrient which is enough for at least 97% of the population.

Nutrient Intakes for Zinc, mg/day.

Age	RNI	.	Age	RNI
0 to 6 months	4.0mg	.	Adult men	9.5mg
7 months to 3 yrs	5.0mg	.	Adult women	7.0mg
4 to 6 yrs	6.5mg	.	.	.
7 to 10 yrs	7.0mg	.	Pregnant women	7..0mg
11 to 14 yrs	9.0mg	.	Breast feeding women	9.5 to 13mg

Pregnant women do need extra zinc, but it is thought that demands are met by increased absorption from the gut. Breast feeding mothers need extra zinc in their diet. Breast milk contains over 2mg of zinc in the first four months and women need an extra 6mg a day to cover these demands. Breast feeding women must include a generous serving of at least one good source of zinc in their diet each day. The iron supplements routinely prescribed for pregnant women when there is no sign of anaemia may compromise zinc status.

This is because the interactions between essential minerals are complex and too much of one may cause an imbalance of another. Excess zinc is toxic. Too much zinc will interfere with the metabolism of other minerals in the body, particularly iron and copper. Symptoms of zinc toxicity occur after ingestion of 2g of or more and include nausea, vomiting and fever.

Meal Plan

Sample one day's meal plan to meet the RNI of 9.5mg for an adult man.					
Breakfast		**Lunch**		**Evening meal**	
Bowl of muesli with milk	1.32mg	Baked beans (225g or 8oz)	1.57mg	Potatoes, boiled (200g or 7oz)	0.4mg
1 slice of toast	0.8mg	2 slices of toast	1.6mg	Peas (80g or 2.5oz)	0.56mg
.	.	Cheese (30g or 1oz) (or Pumpkin seeds, 20g or 2/3oz)	1.2mg	1 Vegeburger (100g or 3.5oz)	1.3mg
.	.	.	.	1 Fruit yoghurt	0.94mg
Total zinc intake 9.69mg					

ALCOHOL

The use of animal derived products in the production of alcoholic beverages is fairly widespread not because no alternatives exist, but because they always have been used and there is little demand from the consumer for an alternative.

The main obstacle when trying to judge the acceptability to vegetarians of any given product is a clause in the 1984 Food Labelling Regulations (UK) which exclude from the 1984 Food Act all drinks with an alcohol content exceeding 1.2% by volume (ABV), leaving only very low or non-alcoholic beers, wines and ciders being required to list all ingredients.

The main appearance of animal derived products is in the fining or clearing process, though others may be used as colorants or anti-foaming agents. It must be pointed out that alcohol is routinely tested on thousands of animals each year (though this is not usually done directly by any individual company).

Beer

Cask-conditioned ales need fining to clear the material (especially the yeast) held in suspension in the liquid. This is invariably done by adding isinglass, derived from the swim bladders of certain tropical fish especially the Chinese sturgeon, which acts as a falling suspension.

If you were to hold a pint of real ale up to the light and see cloudy lumps swirling around that would suggest that the cask had been recently disturbed and the isinglass shaken up from the bottom. Bottled naturally conditioned beers will not always have been treated with isinglass. Keg Beers and Lagers are pasteurised and usually passed through Chill Filters, as are canned beers and some bottled beers, however a considerable number of breweries still use isinglass to clear their pasturised beers, though sometimes only to rescue selected batches which are considered too hazey. Also occasionally the sometimes animal derived additive Glyceryl Monostearate is used in place of 900 Dimethylpolysiloxane as a foam-control agent in the production of keg beers.

It is sometimes possible to buy barrels of cask-conditioned beer from a brewery before it has been fined. The beer would then have to be left for a considerable time to stand before consumption. To our knowledge, only one pub in England sells unfined real ale on draught: The Cumberland Arms in Byker, Newcastle on Tyne.

Cider

Most of the main brands of cider will have been fined using gelatine. Scrumpy type ciders are less likely to have been fined.

Wine

With wine, it is again in the fining process that animal derived ingredients make an appearance. Finings can be isinglass, gelatin, egg albumen, modified casein (from milk), chitin (derived from the shells of crabs or lobsters) or ox blood (rarely used today). But alternatives do exist in the form of bentonite, kieselguhr, kaolin and siica gel or solution. Also newer methods such as centrefuging and filtering are becoming more popular.

The majority of organic wines do not use animal derived finings-but some do, Thorson's Organic Wine Guide by Jerry Lockspeiser and Jackie Gear, lists those wines which are suitable. You might like to note that the Wine Development Board claim that the fining agents are removed at the end of the process with the possible exception of very minute quantities.

Spirits

Most spirits appear to be acceptable to vegetarians, with the possible exception of Malt Whisky, some blended whiskies and

Spanish Brandies which have been conditioned in casks which had previously held sherry which may have been treated with animal derived finings. (Brandy itself is not produced from wine which has undergone any fining processes). Also some imported Vodkas may have been passed through a bone charcoal filter.

Fortified Wines

All ports except crusted port are fined using gelatin. Sherry should be treated in a similar way to wine.

Colorants

E120 cochineal produced by extracting the red body material from pregnant scale insects of the species Dactilopius Coccus is used as a colorant in a small number of red wines, soft drinks and Campari.

CHEESE AND RENNET

Cheese is made by coagulating milk to give curds which are then separated from the liquid, whey, after which they can be processed and matured to produce a wide variety of cheeses. Milk is coagulated by the addition of rennet. The active ingredient of rennet is the enzyme, chymosin (also known as rennin). The usual source of rennet is the stomach of slaughtered newly-born calves. Vegetarian cheeses are manufactured using rennet from either fungal or bacterial sources. Advances in genetic engineering processes means they may now also be made using chymosin produced by genetically altered micro-organisms.

Manufacture

The exact processes in the making of cheese varies between different varieties. However, all cheeses are made by essentially the same method. Initially, the milk is usually pasteurised by heating at 72°C for 15 seconds to destroy potentially harmful bacteria. The milk is then cooled to around 30°C and a starter culture of lactic acid bacteria is added to help souring. These convert lactose into lactic acid and help in the coagulation process. In addition, they also have a beneficial effect on the eventual quality, taste and consistency of the cheese. Some cheeses are coagulated entirely by lactic acid bacteria and are known as lactic-curd or acid-curd cheeses. However, some cheeses sold as lactic-curd cheese may have had rennet added.

The next stage is the addition of rennet, containing the enzyme chymosin. Rennet is usually sourced from the abomasum (fourth stomach) of newly-born calves. Here, chymosin aids the digestion and absorption of milk. Adult cows do not have this enzyme. Chymosin is extracted by washing and drying the stomach lining, which is then cut into small pieces and macerated in a solution of boric acid or brine at 30°C for 4-5 days. Pepsin may sometimes be used instead of chymosin. This is usually derived from the abomasum of grown calves or heifers, or less commonly pigs. Pepsin may be mixed with calf rennin. Rennet coagulates the milk, separating it into curds and whey. This is called curdling.

Chymosin breaks down the milk protein casein to paracasein which combines with calcium to form calcium paracaseinate, which separates out. Milk fat and some water also becomes incorporated into this mass, forming curds. The remaining liquid is the whey. The strength of different rennets can vary, though usual strength varies between 1:10,000 and 1:15,000 i.e. one part rennin can coagulate 10-15,000 parts milk.

Other substances may also be added during the cheese making process. Calcium chloride is added to improve the curdling process, and potassium nitrate is added to inhibit contaminating bacteria. Dyes (e.g. annatto, beta-carotene), Penicillium roquefortii mould spores to promote blue veining, or propionic acid bacteria to encourage hole formation may be added.

Following curdling, the curds are cut and drained. The size of the cut and the methods used vary for different cheese varieties. For soft cheeses, the curds are sparingly cut and allowed to drain naturally. For hard cheeses, the curds are heated and more whey is drained off. The curds are then cut into small pieces, placed in vats and pressed.

After pressing, the curds may be treated in a number of ways. They may be moulded into different shapes, soaked in a saltwater solution, be sprayed with mould forming spores or bacteria, washed in alcohol, or covered in herbs.

The final stage is ripening, or maturation. This can vary in length from 4 weeks to 2-3 years, depending on the type of cheese. During ripening flavours develop, the cheese becomes firmer and drier, and special characteristics such as holes, blue veining and crust formation occurs.

Vegetarian Cheeses

Vegetarian cheeses are made with rennets of non-animal origin. In the past, fig leaves, melon, wild thistle and safflower have all supplied plant rennets for cheese making. However, most widely available vegetarian cheeses are made using rennet produced by fermentation of the fungus Mucor miehei. Vegetarian cheese may also be made using a rennet from the bacteria Bacillus subtilis or Bacillus prodigiosum.

Advances in genetic engineering techniques mean that some vegetarian cheeses may now be made using chymosin produced by genetically engineered micro-organisms. The genetic material (DNA) which encodes for chymosin is introduced into a micro-organism which can then be cultured to produce commercial quantities of chymosin.

This is done by extracting genetic material from calf stomach cells which acts as a template for producing the chymosin encoding DNA. This can then be introduced into the micro-organism. Once the genetic material is introduced there is no further need for calf cells. Alternatively, the chymosin encoding DNA can be bio-synthesised in the laboratory without the use of calf cells.

The chymosin produced is identical to that produced by calf stomach cells. The development of genetically engineered chymosin has been encouraged by shortages and fluctuations in cost of rennet from calves. It's manufacturers claim that genetically engineered chymosin will end the cheese making industry's reliance on the slaughter of calves.

Chymosin encoding DNA has been introduced into three different micro-organisms. These are the yeast Kluyveromyces lactis, the fungus Aspergillus niger var awamori, and a strain of the bacteria Escherichia coli. All of these have now been approved and cleared for use by the Ministry of Agriculture, Fisheries & Food. There is no legal requirement for manufacturers to state whether a genetically engineered rennet has been used in the cheese making process.

Vegetarian cheeses are widely available in supermarkets and health food stores. A wide variety of cheeses are now made with non-animal rennet and labelled as suitable for vegetarians. No particular type of cheese is exclusively vegetarian. Soft cheeses are as likely to be non-vegetarian as hard cheese.

Types of Cheese

The type of cheese produced depends on the milk used and the cheese making process. The milk used may be full fat, semi-skimmed or fully skimmed, this affecting the fat content of the cheese. It may be pasteurised or unpasteurised. Milk from different animals and different breeds is important in determining the final flavour. As well as cow's milk, cheese may be made from sheep or goat's milk.

Soft cheeses may be fresh or ripened. Fresh cheeses include quark, cottage cheese and cream cheese. Ripened soft cheeses include Brie and Camembert. Semi-soft cheeses include Stilton, Wensleydale and Gorgonzola. Hard cheeses include Cheddar, Cheshire, and Gruyere. Parmesan is a strongly pressed, very hard, dry cheese ripened for 2-3 years and then grated. Whey cheeses such as Ricotta are made as a by-product of other cheeses from the whey removed during pressing. Processed cheeses are either made with trimmings that are left over from the manufacture of other cheeses, or from dried milk powder. Flavourings, colourings and other additives are used.

Nutritional Aspects

Cheese is a good source of protein, calcium, zinc, and vitamin B12. However, full fat cheese is a major source of saturated fat which can lead to raised serum cholesterol levels. Also, it contains no carbohydrate or fibre, and is a very poor source of iron.

Vegetarians, particularly new vegetarians, should be wary of too high a consumption of cheese.

NUTS AND SEEDS

Nuts are seeds that are covered with a hard shell. Most are the seeds of trees, but the seeds of a few other plants that are not strictly nuts will also be considered here as they can be conveniently classified with nuts for culinary purposes.

Nuts can be used in many ways. Whole, flaked and ground nuts and nut butters are widely available. A classic vegetarian savoury is nut roast and many vegetarian cook books give a recipe for one, which can be endlessly varied with different herbs and flavourings and different combinations of nuts and cereals. Nuts can be added to sweet dishes, cakes and biscuits, and nut butters can be added to soups and stews to thicken them.

Nutrition

Nuts in general are very nutritious, providing protein and many essential vitamins, such as A and E, minerals, such as phosphorous and potassium, and fibre.

Nuts are also high in carbohydrate and oils, so shouldn't be eaten in excess. Whereas pulses all belong to the legume group of plants, nuts come from a variety of different plant groups, so the nutritional content is more varied too. A brief description of individual varieties is given below, together with the main nutrients they contain.

Storage

Nuts should be stored in cool, dry conditions in airtight containers away from the light. Because of their high fat content, many of them benefit from storage in the fridge or freezer to deter rancidity.

NUTS

Almonds

Probably originated in the Near East but now grows in Southern Europe, Western Asia, California, South Australia and

South Africa. Almond oil is used for flavouring and for skin care preparations and is extracted from the kernel of the Bitter Almond. The Sweet Almond is grown for nuts for eating and have the largest share of the nut trade world-wide. Almond flour is available and it is possible to make a nutritious nut milk from almonds. Almonds are particularly nutritious, 100g contain 16.9g protein, 4.2mg iron, 250mg calcium, 20mg vitamin E, 3.1mg zinc and 0.92mg vitamin B2.

Brazils

A native of South America. The nuts grow inside a hard, woody fruit rather like a coconut shell which has to be broken open to expose the 12-24 nuts inside. Brazils are high in fat, which causes them to go rancid very quickly, and protein. 100g of brazils contain 12g protein, 61g fat, 2.8mg iron, 180mg calcium, 4.2mg zinc.

Cashews

Native to America but now grown extensively in India and East Africa. It will withstand rather drier conditions than most other nuts. The nut grows in a curious way on the tree, hanging below a fleshy, apple-like fruit. It is related to the mango, pistachio and poison ivy. High in protein and carbohydrate, 100g cashews contain 17.2g protein, 60 micrograms vitamin A, 3.8mg iron.

Chestnuts

The sweet chestnut is a native of South Europe but is planted eisewhere extensively for both nuts and timber. The nuts can be used in soups, fritters, porridges, stuffings and stews, as well as being roasted or boiled whole.

Available fresh (in autumn), dried, canned-whole or pureed, or ground into flour. Dried chestnuts need soaking for at least 1-2 hours and boiling for 45-60 minutes, fresh need boiling for 40 minutes before being peeled. Preserved in syrup they become the famous delicacy, Marron-glace. High in starch, but low in protein and fat, 100g chestnuts contain 36.6g carbohydrate, only 2g protein (the lowest of all nuts) and 2.7g fat.

Coconuts

The coconut palm is common in tropical regions all over the world. The nut is covered in a fibrous outer coating on the tree and all parts of the tree are useful, the trunks for timber, the leaves for thatch, the fibrous husk produces coir-the starting material for ropes and coconut matting-and the nuts are used for food. Unripe nuts contain coconut milk. The nutmeat can be eaten fresh or dried (desiccated or flaked coconut) and is also available in blocks of creamed coconut. A valuable oil is also extracted from the nut meat and used for cooking (although it is very high in saturated fat), margarines, soaps and detergents. 100g fresh coconut contain 3.2g protein and 36g fat, dessicated contain 5.6g protein and 62g fat.

Hazels

Hazel, also called Cob, is a common wild tree in Europe and Asia and its nuts have been eaten by humans since earliest times. The cultivated varieties are bigger and the filbert is a similar but bigger species from SE Europe. Used in sweet and savoury dishes, they are available whole, ground and flaked, or made into oil and nut butter. 100g hazel nuts contain 7.6g protein, and they are lower in fat than most other nuts.

Macadamia Nuts

A native of NE Australia now also grown commercially in Hawaii. Notoriously difficult to extract from their shells, they are expensive but have a delicious creamy flavour and crunchy texture. Low in carbohydrate, but quite high in fat, 100g Macadamia nuts contain 7g protein and 40mg calcium.

Peanuts

Also known as groundnuts or monkey nuts, peanuts are actually legumes. Of South American origin, it's now an important crop all over the tropics and southern USA. It gets its name groundnut because as the pods ripen, they are actually forced underground. Peanuts are high in protein and contain 40-50% oil. The oil is used in cooking, as salad oil, in margarines and the residue is fed to animals. Whole peanuts can be eaten raw or roasted or made into peanut butter (look out for brands which do not contain hydrogenated oils, which are highly saturated). As they are usually inexpensive, they can be mixed with other kinds of nuts to bring down the cost, while still maintaining flavour and good nutrition. 100g peanuts contain 24.3g protein, 2mg iron and 3mg zinc.

Pecans

A native of N America where it is used extensively in ice cream, cakes, nut bread and confectionery. The flavour is rather like a mild, sweet walnut. 100g pecans contain 9.2g protein, a very high fat content of 71.2g, 130 micrograms vitamin A (also very high), 2.4mg iron and 73mg calcium.

Pine Nuts

These are the seeds of the Stone Pine, a native of the Mediterranean region, but the seeds of various other pines are eaten in various parts of the world including the seeds of the Korean Pine or North American pinon tree. They are very difficult to harvest, hence their cost. They are vital for pesto sauce, and are delicious lightly toasted. They become rancid very easily and should be stored in the fridge or freezer. 100g pine nuts contain 31g protein, the highest of the nuts and seeds.

Pistachios

Native to the Near East and Central Asia but has long been cultivated in the Mediterranean region and more recently in the Southern US. The kernels are green and are prized as much for their ornamental colour as for their flavour. Also sold roasted and salted in their shells. They are more expensive than most other nuts. 100g pistachios contain 19.3g protein, 14mg iron, 140mg calcium.

Walnuts

The walnut is native to SE Europe and West & Central Asia but is now grown in the UK, California and China as well. It is grown for timber as well as its nuts. Walnut oil has been used for centuries in the preparation of artists paints. The black walnut is a native of North America, introduced into Britain in the 17th century. The butternut is also from North America. These two have much thicker shells than European walnuts. High in fat, they go rancid very quickly and should be stored in the fridge or freezer. 100g walnuts contain 10.6g protein and 2.4mg iron.

SEEDS

Pumpkin

Can be eaten raw or cooked in both sweet or savoury dishes. Delicious toasted and sprinkled, while hot, with soya sauce and served on salads. They are rich in protein, iron, zinc and phosphorous. 100g pumpkin seeds contain 29g protein, 11.2mg iron and 1144mg phosphorous.

Sesame

Of African origin but now common in tropical and sub-tropical Asia. An oil is extracted from the seed and used for cooking, salad oil and margarines. It is also available as toasted sesame oil for oriental cooking. The whole seeds can also be eaten and are most often seen as a decoration on cakes, confectionery etc. Sesame seed paste, tahini, is used in many dishes e.g. hummus. Halva, a sweet made from sesame seeds is often found in health food shops. A good source of protein and calcium, 100g sesame seeds contain 26.4g protein, 12.6mg vitamin B3, 7.8mg iron, 131mg calcium and 10.3mg zinc.

Sunflower

An annual plant belonging to the daisy family, it probably originated in North America or Mexico. North American Indians cultivated sunflowers as long as 2,000 years ago. The oil extracted from its seeds is used in margarine, varnishes and soaps but the seeds can be eaten whole, raw or cooked. They can be added to breads and cakes or sprinkled over salad or breakfast cereals. A good source of potassium and phosphorous, 100g sunflower seeds also contain 24g protein and 7.1mg iron and 120mg calcium.

PULSES

Peas, beans and lentils are known as pulses. They are the seeds of plants belonging to the family Leguminosae, which gets its name from the characteristic pod or legume that protects the seeds while they are forming and ripening. With approximately 13,000 species, the family Leguminosae is the second largest in the plant kingdom and it is very important economically.

Different kinds of legumes provide us with food, medicines, oils, chemicals, timber, dyes and ornamental garden plants. Legume products include carob, senna, gum arabic, balsam, indigo and licorice. Pulses are valuable because they contain a higher percentage of protein than most other plant foods.

Origins

Pulses have been used as food for thousands of years. The lentil was probably one of the first plants ever to be domesticated by humans. Most pulses prefer warm climates but there are varieties which grow in temperate regions. They can be eaten fresh or dried and come in a great number of varieties with a range of colours, flavours, and textures. In spite of its common name, the peanut or groundnut is also a legume rather than a nut.

Nutrition

All pulses, except for soya beans, are very similar in nutritional content. They are rich in protein, carbohydrate and fibre, and low in fat which is mostly of the unsaturated kind. They are also important sources of some B vitamins. Fresh pulses contain vitamin C, but this declines after harvesting and virtually all is lost from dried pulses. Canned pulses however, retain about half their vitamin C except for canned, processed peas which have been dried before canning.

Canning doesn't affect the protein content, eliminates the need for soaking and considerably reduces the cooking time compared with dried pulses. Frozen peas will have also lost about a quarter of their vitamin C content. Pulses are usually eaten for their high protein content. A typical nutritional breakdown is that for haricot beans which are used to make baked beans, contain, per 100g dried beans: 21.4g protein, 1.6g fat, 45.5g carbohydrate, 25.4g fibre, 6.7mg iron and 180mg calcium.

The nutritional quality of the soya bean is superior to that of other pulses. It contains more protein and is also a good source of iron and calcium. The nutritional breakdown of soya is per 100g of dried beans: 34.1g protein, 17.7g fat, 28.6g carbohydrate, 8.4mg iron and 226mg calcium. Dried soya beans are lengthy to prepare because they need at least 12 hours soaking and 4 hours cooking time, boiling for the first hour, but nowadays a large number of soya based foods including tofu, tempeh and textured vegetable protein (soya mince or chunks) are available.

Storage and Cooking

One advantage of dried pulses is that they will store very well for long periods if kept in a dry, airtight container away from the light. However it is best to eat them as fresh as possible. Pulses toughen on storage and older ones will take longer to cook. Allow about 55g dried weight per person, once soaked and cooked they will at least double in weight. Most dried pulses need soaking for several hours before they can be cooked, exceptions are all lentils, green and yellow split peas, blackeye and mung beans.

Soaking times vary from 4-12 hours, it is usually most convenient to soak pulses overnight. Always discard the soaking water, rinse and cook in fresh water without any salt, which toughens the skins and makes for longer cooking. Changing the water will help to reduce the flatulence some people suffer when eating pulses, also reputed to help is the addition of a pinch of aniseeds, caraway, dill or fennel seeds.

Toxins in Pulses

Consumers should be aware that it is not safe to eat raw or undercooked kidney and soya beans. There is no need to avoid them as long as they are thoroughly cooked.

Red kidney beans: Incidents of food poisoning have been reported associated with the consumption of raw or undercooked red kidney beans. Symptoms may develop after eating only four raw beans and include nausea, vomiting and abdominal pain followed by diarrhoea. A naturally occurring haemaglutin is responsible for the illness, but can be destroyed by high temperature cooking, making the beans completely safe to eat.

For this reason, kidney beans must not be sprouted. Kidney beans should be soaked for at least 8 hours in enough cold water to keep them covered. After soaking, drain and rinse the beans, discarding the soaking water. Put them into a pan with cold water to cover and bring to the boil. The beans must now boil for 10 minutes to destroy the toxin. After this the beans should be simmered until cooked (approximately 45-60 minutes) and they should have an even creamy texture throughout-if the centre is still hard and white, they require longer cooking.

Soya Beans: Contain an anti-trypsin factor (or trypsin inhibitor) which prevents the assimilation of the amino acid methionine. Soya beans also require careful cooking to ensure destruction of this factor. They should be soaked for at least 12 hours, drained and rinsed then covered with fresh water and brought to the boil. Soya beans should be boiled for the first hour of cooking. They can then be simmered for the remaining 2-3 hours that it takes to cook them. Soya flour should state heat treated on its packaging. Other soya products (e.g. tofu, tempeh, soya milk, soya sauces and miso) are quite safe to use. Soya beans can be sprouted, but the sprouts should be quickly blanched in boiling water to inactivate the trypsin inhibitor.

Pressure Cooking: The temperatures achieved in pressure cooking are adequate to destroy both haemaglutins and the trypsin inhibitor. Pressure cooking also considerably reduces cooking times-kidney beans 10-20 minutes, soya beans 1 hour.

Canning: The temperature achieved in the canning process also renders pulses quite safe.

Slow Cookers: Pulses must be soaked and boiled for 10 minutes before being added to a slow cooker, as they do not reach sufficiently high temperatures to destroy the toxins. As beans and peas are all very similar nutritionally, with the exception of soya, they can be interchanged in most recipes if you want to experiment or have run out of one kind, as long as you take into account the different cooking times. If the beans are likely to need a lot longer to cook than the other ingredients, try pre-cooking them in a separate pan before adding to the other ingredients or using canned beans.

Sprouting

Many whole pulses (e.g. aduki, chickpeas, whole lentils, marrowfat peas, mung and soya beans) can be sprouted which increases their nutritional value.

SOYA AND MYCOPROTEIN

Soya

The soya bean is the seed of the leguminous soya bean plant. Soya foods have been a staple part of the Chinese diet for over 4000 years but have only been widely consumed in Western countries since the 1960's. Soya foods include tofu, tempeh, textured vegetable protein (chunks, mince etc), miso, soya sauces, soya oil and margarine, and soya dairy alternatives.

Soya is an excellent source of high quality protein, is low in saturated fats and is cholesterol free. Recent research has indicated soya has several beneficial effects on health in addition to its nutritional benefits. Soya beans contain high concentrations of several compounds which have demonstrated anti-carcinogenic activity. These include isoflavonoids, protease inhibitors and phytic acid. The low incidence of breast and colon cancer in China and Japan has been partially attributed to the high consumption of soya products. The low incidence of menopausal symptoms in Japanese women has also been attributed to high consumption of soya. Soya diets have also been shown to reduce levels of serum cholesterol.

Textured Vegetable Protein

Textured vegetable protein is basically defatted soya flour which has been processed and dried to give a substance with a sponge-like texture which may be flavoured to resemble meat. Soya beans are dehulled and their oil extracted before being ground into flour. This flour is then mixed with water to remove soluble carbohydrate and the residue is textured by either spinning or extrusion. Extrusion involves passing heated soya residue from a high pressure area to a reduced pressure area through a nozzle resulting in the soya protein expanding. The soya protein is then dehydrated and may be either cut into small chunks or ground into granules. Textured vegetable protein may be purchased either unflavoured or flavoured to resemble meat. It is prepared simply by mixing with water or stock and leaving to stand for a few minutes, after which it may be incorporated into recipes as a meat substitute. Soya protein is also available incorporated into various vegetarian burgers, sausages, canned foods etc. As well as being a good source of fibre and high quality protein, Textured vegetable protein is fortified with vitamin-B12.

Tofu

Tofu is soya bean curd made from coagulated soya milk. Soya beans are soaked, crushed and heated to produce soya milk to which a coagulating agent such as calcium sulphate or calcium chloride is added. The resulting soya curd is then pressed to give tofu. Tofu is sometimes known as soya cheese, and is sold as blocks packaged in water. It can be bought as silken tofu, which is soft and creamy in texture, or as a denser, firmer version. The firmer kind may also be purchased smoked or marinated. Tofu tends be fairly bland tasting and is best used in recipes where flavour is imparted by other ingredients. Firm tofu may be marinated, fried, stir-fried, deep-fried, sauteed, diced and added to salads or casseroles. Silken tofu can be used for dips, spreads, sauces and sweet dishes. As well as having a high protein content, tofu also contains calcium, iron, and vitamins B1, B2 and B3.

Tempeh

Tempeh is a fermented soya bean paste made by inoculating cooked soya beans with the mould Rhizopus oligosporous. This mould forms a mycelium holding the soya beans together and is responsible for the black specks in tempeh. Tempeh has a chewy texture and distinctive flavour and can be used as a meat substitute in recipes. It may be deep-fried, shallow-fried, baked or steamed.

Miso

Miso is a fermented condiment made from soya beans, grain (rice or barley), salt and water. Miso production involves steaming polished rice which is then inoculated with the fungus Aspergillus oryzae and left to ferment to give an end product called koji. Koji is then mixed with soya beans which have been heated and extruded to form strands, together with salt and water. This is then left to ferment in large vats. Miso varies widely in flavour, colour, texture and aroma. It is used to give flavour to soups, stews, casseroles, and sauces.

Soya Sauces

True soya sauce, called shoyu, is made by fermenting soya beans with cracked roasted wheat, salt and water. Tamari is similar but slightly stronger and made without wheat (and so is gluten-free). Fermentation for shoyu and tamari takes about one year. Much of the soya sauce available in supermarkets is not true soya sauce but is made by chemical hydrolysis from defatted soya flour, caramel colouring, and corn syrup without any fermentation process.

Soya Dairy Alternatives

Soya milk is an alternative to dairy milk and is widely available in supermarkets and health food stores. It is most commonly made by soaking soya beans in water which are then strained to remove the fibre. It can also be made from soya protein isolate or soya flour. Compared to full fat cow's milk, soya milk has a lower fat content, a lower proportion of saturated fat, and no cholesterol. It is low in carbohydrate and provides a good source of protein. Some brands may be fortified with calcium, vitamin-D2, vitamin-B12 and vitamin-B2. Soya milk provides an alternative to cow's milk for people with cow's milk protein and lactose intolerance. Cow's milk allergy is most common in infants, and specially formulated soya milks are available for babies. Other soya milks are not suitable as sole foods for young infants.

Previously, the media has linked soya milk with having a high aluminium content. However, the aluminium content of soya milks is generally lower than cow's milk, and falls well within acceptable limits dictated by the World Health Organisation. Aluminium in soya milks can be regarded as negligible. Certain infant formulas (both cow's milk and soya milk based) produced from concentrates have been reported as having high levels of aluminium and their suitability for infants has been questioned. A number of different brands of soya milk may be purchased. These may be sweetened or unsweetened and vary in flavour. Market leaders are Provamel, Granose and Plamil. Some supermarkets also sell own-brand soya milk. In addition to soya milk, a range of flavoured soya desert and soya yoghurt products are available.

Other Soya Products

Soya oil and margarine are widely used and are high in polyunsaturated fats and low in saturated fats. Other less easily available soya foods include soya sprouts, soya nuts (roasted and seasoned soya beans), natto (fermented soya beans made with a bacteria, Bacillus subtilis), yuba (the skin formed on heated soya milk), soya flakes, soya flour, and high protein soya isolates and concentrates.

Mycoprotein

Mycoprotein is a food made by continuous fermentation of the fungus, Fusarium gramineurum. The fungus is grown in a large fermentation tower to which oxygen, nitrogen, glucose, minerals, and vitamins are continually added. After harvesting, the fungus is heat treated to reduce its RNA content to World Health Organisation recommended levels before being filtered and drained. The resulting sheet of fungal mycelia is mixed with free range egg egg albumen which acts a binder. Flavouring and colouring may also be added. The mycoprotein is then textured to resemble meat, before being sliced, diced or shredded. Mycoprotein is a source of protein, fibre, biotin, iron and zinc, and is low in saturated fat. Mycoprotein was developed by Rank Hovis McDougall, and is marketed under the name of Quorn by Marlow Foods Ltd (now owned by Premier Foods). A wide range of Quorn ready meals are available including curries, pies, and casseroles, and it may also be purchased as chunks, mince, sausages, burgers, fillets etc. Since January 2005 the entire Quorn range have been approved by the Vegetarian Society (previously the manufacturers had not been able to source enough free range eggs to use accross whole product range).

Wheat Protein

Wheat protein is derived from wheat gluten. It is sometimes called Seitan. Gluten is extracted from wheat and then processed to resemble meat. Wheat protein is marketed under the name of Wheatpro by Lucas Ingredients of Bristol. It has a greater similarity to meat than textured vegetable protein or mycoprotein and is used as a meat substitute in a range of foods. It is available in some health food stores.

Stumbling Blocks

- *Additives.*
- *Albumen:* Derived from eggs, probably battery.
- *Alcohol:* Many alcoholic drinks are fined (ie clarified) using animal ingredients, see beer and wine. Spirits are suitable for vegetarians except for some Russian and Eastern European Vodkas which may have used bone charcoal in their production. Watch out for cochineal in Campari.
- *Alpaca:* Animal derived clothing material.
- *Anchovies:* Small fish, found on pizzas and in some brands of worcester sauce.
- *Angora:* Animal derived clothing material.
- *Animal:* Fat Carcass fat not milk fat.
- *Aspic:* Savoury jelly derived from meat or fish.
- *Beer:* All cask conditioned "real" ales will have been fined with isinglass, and some keg, bottled and canned bitters, milds and stouts also. Lagers are generally chill filtered, but some brands may use isinglass on occasion.
- *Biscuits:* Quite likely to contain animal fats.
- *Bone:* Used in bone china and cutlery handles.

- *Bread:* Most large producers use vegetable based emulsifiers (E471, E472 etc), but local bakers may not. Some bakers may grease the tins with animal fat.
- *Breakfast Cereals:* Often fortified with vitamin D3.
- *Brushes:* Animal hair is commonly used for paint and shaving brushes.
- *Butter:* Pure butter is suitable for vegetarians.
- *Capsules:* Usually made from gelatine, vegetarian alternatives are coming onto the market.
- *Cashmere:* Animal derived clothing material.
- *Catering/Cookery:* Training may require the handling of meat.
- *Caviar Fish Eggs:* The fish must be killed to obtain the eggs.
- *Cheese:* Likely to have been produced using animal rennet.
- *Chewing Gum:* Often contain glycerine. Wrigleys use a vegetable glycerine.
- *Chips:* May have been fried in animal fat.
- *Chitin:* Produced from crab & shrimp shells.
- *Chocolate:* Watch out for whey and emulsifiers.
- *Clothing:* Many materials derived from animals, others causing environmental problems.
- *Cochineal E120,* made from crushed insects.
- *Crisps:* Often use whey as a flavour carrier, ready salted are the only clearly vegetarian flavour, though some beef crisps are flavoured with yeast extract and are therefore suitable.
- *Down:* Usually from slaughtered ducks or geese, though some live plucking does occur, used in bedding.
- *E Numbers:* European food additives numbering system, not all vegetarian.
- *Edible Fats:* Can mean animal fats.
- *Eggs:* Some vegetarians may wish to avoid battery eggs and/or barn eggs. The Vegetarian Society does not award

its seedling symbol to any products containing eggs other than free range.

- *Emulsifiers:* May not be vegetarian.
- *Fast Food:* Watch out for Bean/Vegetable burgers being cooked with fish/chicken/meat products.
- *Fatty Acids:* May be of animal or vegetable origin.
- *Feathers:* Clothing material
- *Felt:* Made from wool or fur.
- *Fur:* Clothing material
- *Gelatin/Gelatine:* A gelling agent derived from animal ligaments, skins, tendons, bones etc. Alternatives such as Agar Agar, Carrageen and Gelozone exist.
- *Glycerine/Glycerol:* May be produced from animal fats, synthesised from propylene or from fermentation of sugars.
- *Gravy:* Vegetarian gravy mixes are available. Be careful in restaurants.
- *Honey:* Avoided by most vegans.
- *Ice Cream:* Look out for non dairy fats, E numbers, eggs.
- *Isinglass:* A fining agent derived from the swim bladders of certain tropical fish, especially the Chinese sturgeon.
- *Jelly:* Usually contains gelatine though Alternatives are available.
- *Lactose:* Produced from milk, sometimes as a by product of the cheese making process.
- *Lanolin:* Produced from sheep's wool. Used to make vitamin D3.
- *Leather:* Around 10% of the value of an animal at slaughter is in its skin.
- *Lecithin:* Nearly always produced from soya beans, though can be produced from eggs.
- *Margarines:* May contain animal fats, fish oils, vitamin D3, E numbers, whey, gelatine.
- *Mohair:* Animal derived clothing material.

- *Olive Oil:* No problems! Just worth knowing about.
- *Pasta:* May contain egg.
- *Pastry:* May contain animal fat.
- *Pepsin:* Enzyme from a pig's stomach, used like rennet.
- *Pet Foods:* Dogs are omnivorous and can be fed on an exclusively vegetarian diet. Canned and dried dog foods are available.
- *Photography:* All Photographic film uses gelatine.
- *Postage Stamps:* The backing glue is free from animal products.
- *Rennet:* An enzyme taken from the stomach of a newly killed calf used in the cheese making process. Vegetarian cheese is produced using microbial or fungal enzymes.
- *Restaurants:* Watch out for non-vegetarian cheese, battery eggs, stock.
- *Roe Fish:* Eggs.
- *Shellac:* Secreted under tree bark by insects. To be treated in a similar way to Honey.
- *Shoes:* Quality synthetic shoes are becoming more widely available.
- *Silk:* Harvesting silk used in invariably causes the death of the silk worm.
- *Soap:* Many soaps are not vegetarian since they use animal fats and/or glycerine. Vegetable oil based soaps are quite widely available.
- *Soft Drinks:* Some canned Orange drinks use gelatine as a carrier for added Beta Caratine. (This would not appear on the ingredients panel).
- *Soup:* Watch out for the stock.
- *Spirits:* (alcoholic that is!) possible problems with fining and filtering.
- *Stearic:* Acid May be vegetarian or not.
- *Stock:* May contain animal fat.

- *Suet:* Usually made from animal fat, vegetable versions are available.
- *Sweets:* Look out for gelatine in boiled sweets and mints, and cochineal in boiled sweets and Smarties. (some vegetarian sweets are listed by chocolate manufacturers.)
- *Toothpaste:* Many brands contain glycerine.
- *Vegan:* The Vegan Society produces The Animal Free Shopper which lists branded products suitable for vegans (available from The Vegetarian Society).
- *Vitamins:* Vitamin D2 is produced by sunlight acting on bacteria, however D3 is derived from lanolin from sheeps' wool therefore only D3 which is guaranteed sourced from wool sheared from live sheep is considered acceptable.
- *Seedling Symbol:* You can be sure that any products carrying the Vegetarian Society's V symbol have been thoroughly checked to ensure they are suitable for vegetarians.
- *Washing Powder:* Soap based powders may contain animal fats.
- *Whey:* Whey and whey powder are usually by-products of the cheese making process which mainly uses animal rennet.
- *Wine:* May have been fined using isinglass, dried blood, egg albumen, gelatine, chitin. Vegetarian alternatives include bentonite, kieselguhr, kaolin and silica gel. Non vintage port is fined with gelatine.
- *Wool* may not be so sheep friendly.
- *Worcester:* Sauce Most brands contain anchovies.
- *Yoghurts:* Some low fat yoghurts contain gelatine.

7

VEGETARIAN FOODS AND ITS VITALITY

INTRODUCTION

One development in Britain's eating habits has dwarfed all others during the course of the last twenty years. Unprecedented in both speed and scale, vegetarianism has transformed the way we look at food, and the way we look at ourselves. The figures speak for themselves, we are discussing a change of diet which:

- had 100,000 adherents in 1945 but which today has over 3 million
- has doubled in size over the last ten years, with 2,000 people swelling its ranks every week.
- has convinced 40 per cent of the British population to cut down on or eliminate the red meat in their diet

Vegetarianism is nothing new, and various forms of meat-free eating have been practiced for thousands of years, but in the USA, many parts of Europe, and particularly here in the United Kingdom, this new diet has taken our dinner plates, our menus and our supermarket shelves by storm. Awareness of vegetarian health and vegetarian issues will play an increasingly important part in the work of anyone involved in food, catering, nutrition and health.

So what is a vegetarian? This may seem a superfluous question given the fact that we have all met vegetarians and have probably all eaten vegetarian food, but with so important a phenomenon, common definitions and terms of reference are invaluable. The Vegetarian Society, the acknowledged expert in this area, defines a vegetarian as: One who eats no fish, flesh or fowl and avoids all by-products of slaughter.

Within vegetarianism there are one or two other definitions which may be helpful, they represent the degrees to which vegetarian eating patterns are adhered to by individuals:

A lacto-ovo-vegetarian eats no fish, flesh or fowl but does consume both dairy products and eggs. Lacto ovo-vegetarians are undoubtedly the majority group amongst vegetarians.

A lacto-vegetarian follows the diet of a lacto-ovo-vegetarian, but he or she will not eat eggs.

A vegan is one who avoids all animal products, i.e. fish, flesh, fowl, eggs and dairy products.

Those who are new to vegetarianism may be curious as to why so many people are making so dramatic a change to their diet. Studies show vegetarians to be motivated by three key factors:

Respect for Animals

As our concern over animal welfare grows, particularly over factory farming systems, many individuals have chosen vegetarianism as a way to avoid inflicting unnecessary suffering upon animals.

Environmental Awareness

The Earth's natural resources and habitats are precious indeed.. Pollution, global warming and deforestation can all be linked to livestock farming. Many vegetarians see their diet changes as an immediate and effective way to halt environmental damage.

Human Health

As this report shows, the health benefits of a vegetarian diet are better documented now than ever. Heart disease, for example, unheard of before this century kills more people than any other single cause. Many are moving away from meat for the sake of their health. Every vegetarian is different and so may have other reasons than those above for excluding meat, but each vegetarian has one thing in common: a thought out and forward thinking attitude to the food on their plate.

VEGETARIAN HEALTH

A vegetarian diet confers a wide range of health benefits. Research has proven that vegetarians suffer less from many of the dieases linked to a modern Western diet: obesity, coronary heart disease, hypertension, type II diabetes, diet-related cancers, diverticular disease, constipation and gall stones (British Medical Association,1986; Dwyer, 1988).

Mortality

Epidemiological studies provide clear evidence that vegetarians have a reduced general and cause-specific mortality (McMichael, 1992).

In an 11-year study of 1,900 vegetarians in Germany, Chang-Claude (1992) found mortality from all causes was reduced by one-half compared with the general population. This was mainly attributable to reduced cardiovascular disease in both men and women. A longer duration of vegetarianism was associated with a lower risk, pointing to a real protective effect of the lifestyle (Chang-Claude, 1993).

A 12-year follow-up study of 6,115 British vegetarians and 5,015 meat-eaters found all cause premature mortality to be 20 per cent lower among the vegetarians after adjusting for the confounding effects of smoking, body mass index and socioeconomic status (Thorogood, 1994). Cancer mortality was 39 per cent lower and ischaemic heart disease 28 per cent lower among the vegetarians. A 21-year study of Californian Seventh-Day Adventists also revealed a significant association between meat consumption and all causes of mortality (Kahn, 1984).

While other factors influencing health such as socioeconomic differences, smoking and physical activity may play a confounding role in these studies, there is sufficient evidence to suggest that a reduced risk of mortality is directly linked to a vegetarian diet.

Cardiovascular Disease

Cardiovascular disease is the main cause of premature mortality in Britain, with most deaths being caused by coronary heart disease.

Vegetarians have a significantly lower incidence of coronary heart disease than meat-eaters. Burr and Butland (1988) found mortality from heart disease was 29 per cent lower in a study of more than 6,000 British vegetarians compared with a control group of health-conscious meat-eaters.

A study of more than 25,000 Seventh-Day Adventists noted a definite dose-related link between meat consumption and heart disease (Snowdon, 1984). Among men aged 45 to 64, those who ate meat daily were three times more likely to die from heart disease than those who did not eat meat.

The degenerative processes leading to heart disease can begin at an early age. Atherosclerosis, the forming of fibrous plaques on arterial walls and one of the first stages in the development of cardiovascular disease, has been observed in young children and adolescents. The Coronary Artery Risk Development in Young Adults (CARDIA) study examined the relationship between diet and health in more than 5,000 young adults aged 18 to 30 (Slattery 1991). Vegetarians were found to have a greatly improved cardiovascular fitness and a lower risk of heart disease.

One way that a vegetarian diet is thought to protect against heart disease is the lower cholesterol levels seen in vegetarians. Raised cholesterol is widely recognised as a primary risk factor for heart disease and studies have consistently demonstrated serum cholesterol levels in vegetarians as being around 10 per cent lower than in non-vegetarians (Burr, 1981; Knuiman and West, 1982; Thorogood, 1990).

This figure may result in a 20 to 50 per cent reduction in the incidence of heart disease (Law, 1994). The lower serum cholesterol levels in vegetarians is largely due to lowered low-density-lipoprotein (LDL) cholesterol. This is the cholesterol fraction associated with heart disease.

Healthy eating advice aimed at reducing the risk of heart disease often encourages the use of lean meat cuts. However, in a comparison between a vegetarian diet and a lean meat diet (Kestin, 1989), the vegetarian diet was shown to lower cholesterol more effectively and so lessen the risk of cardiovascular disease.

The California Lifestyle Heart Trial indicated that a low-fat vegetarian diet together with other lifestyle changes such as increased exercise and stress management can actually reverse the progress of heart disease by reducing cholesterol plaques in arteries (Ornish, 1990). Barnard (1992) has shown a low-fat vegetarian diet can achieve high levels of compliance and acceptability among patients with heart disease.

Obesity

Obesity is a major health problem in the UK and its incidence is increasing in both adults and children. Around 12 per cent of women and 8 per cent of men are clinically obese (Gregory, 1990) and many more are overweight.

Obesity predisposes sufferers to a range of conditions including cardiovascular disease, type II diabetes, and hormone-dependent cancers such as breast cancer. Vegetarians are typically leaner than meat-eaters and obesity is less common in vegetarian populations. Body mass index (BMI) for vegetarians is lower than in meat-eaters and closer to desirable levels (Thorogood, 1989).

Levin (1986a) found the prevalence of obesity to be 5.4 per cent in a vegetarian group compared with 19.5 per cent in non-vegetarians. This may be due to vegetarians' lower energy intake overall, with a lower proportion of energy being supplied by fat.

The established measure for obesity is the Body Mass Index (BMI). This is obtained by dividing a persons weight (kg) by the square of their height (m). Normal weight is defined as a BMI of 20-25. A BMI of 25 to 30 is overweight, over 30 is obese.

Obesity is caused by an individual taking in more energy from food than they are using up in physical activity. Dieting is generally ineffective and repeated weight loss and gain can play a part in osteoporosis and heart disease. A healthy diet based on starchy foods such as bread, potatoes, rice and pasta together with fruits and vegetables and combined with increased physical activity is far more effective in bringing about successful weight loss.

Hypertension

Hypertension, or high blood pressure, can contribute to heart disease, strokes and kidney failure. Both cross sectional populaton studies (Armstrong, 1977; Rouse, 1983) and controlled trials in hypertensive subjects (Margetts, 1986) provide evidence that vegetarians have lower blood pressure than non-vegetarians. The reason for this is unclear. Studies suggest that the effect is not due to changes in any specific nutrient intake but the overall effect of combined nutrient change (Beilin, 1990).

Diabetes

Around one million people in the India suffer from diabetes. Of these, about 75 per cent have type II diabetes (non-insulin dependent). A study found vegetarians had a substantially lower risk of type II diabetes than non-vegetarians. The link between meat consumption and incidence of diabetes remained after contributory factors such as weight, physical activity and other dietary factors were accounted for. Dietary recommendations for diabetics encourage a high intake of complex carbohydrates and fibre and low consumption of total and saturated fat. Vegetarian diets tend to match these recommendations more closely. The relative leanness of vegetarians also protects against diabetes, obesity being a major risk factor.

Gall Stones

Gall stones are composed af cholesterol, bile pigments and calcium salts. They form in the gall bladder and can cause severe pain. A study of more than 750 women found that non-vegetarians were almost twice as likely to develop gall stones as vegetarians.

Diverticular Disease and Bowel Function

Diverticular disease is widespread in the UK. Symptoms include abdominal pain and a disturbed bowel habit, usually constipation but sometimes diarrhoea. Gear (1979) found it to be significantly less frequent among vegetarians, probably due to the higher fibre content of vegetarian diets. The high fibre content of vegetarian diets has also been linked with decreased incidence of constipation, appendicitis, irritable bowel syndrome, haemorrhoids and varicose veins (Dickerson, 1985).

Cancer

Cancer is second only to coronary heart disease as a cause of death in Britain. Accounting for around 25 per cent of all premature deaths with lung, breast and colo-rectal cancers being the most common. It is estimated that at least one third all cancers are directly related to diet. Research has indicated that vegetarian diets may offer protection against diet-related cancer.

Plant Foods and Cancer

Plant foods contain a number of substances which are believed to protect against cancer. Indoles, lignans, isoflavones, protease inhibitors and others have all been shown to be potent anti-carcinogens and may play an important role in the lower cancer incidence among vegetarians. In contrast, cooked meat and fish contains carcinogens known as heterocyclic amines (HA's). These are present at high levels in the urine of people consuming cooked meats and have been shown to be metabolically active in humans. Evidence suggests meat-derived HA's may play a role in breast, colon and pancreatic cancer (Snyderwine 1994).

HEALTHY EATING

Vegetarian diets vary greatly between individuals, as do eating patterns. However, a typical vegetarian diet can closely match current dietary recommendations for healthy eating. There is a high level of consensus concerning healthy eating recommendations. Cannon (1992) analysed one hundred reports on food, nutrition and public health published between 1961 and 1991 and found that experts worldwide agreed that diets high in fat and saturated fat and with insufficient fruit, vegetables, cereals and other starchy foods were a major cause of coronary heart disease, some cancers, obesity, diabetes and constipation. Current healthy eating advice focuses on reducing total and saturated fat intake while increasing the consumption of complex carbohydrates and dietary fibre. The major sources of total fat and saturated fat in the diet are meat and dairy products. Complex carbohydrates and fibre are provided by plant foods. By far the easiest method of successfully implementing dietary advice is to limit the intake of foods of animal origin and increase the consumption of fruit, vegetables, cereals and pulses. As a basic guide, it is recommended that a vegetarian diet should include the following each day:

- 3 or 4 servings of cereals/grains-provides energy, fibre, B vitamins, calcium and iron.

  2 or 3 servings of pulses, nuts or seeds-provides protein, energy, fibre, calcium, iron and zinc
- 4 or 5 servings of fruit or vegetables, including: dark green leafy vegetables-for folate, calcium and iron; red, orange and yellow vegetables-for beta-carotene; fresh fruit-for vitamin C; dried fruit-for fibre and iron.
- 2 servings of dairy or soya products-provides protein, energy, calcium and other minerals, vitamin B12, vitamin D.
- A small amount of plant oils, margarine or butter-provides energy, essential fatty acids, vitamin E (plant oils) and vitamins A and D (margarine or butter).

VEGETARIAN NUTRITION

Energy

Energy intakes of vegetarians and vegans meet UK recommended levels (Thorogood, 1990; Draper, 1993) and are in line with World Health Organisation recommendations for adults engaged in light to moderate activities (WHO, 1985). They are also similar to recorded energy intakes for meat-eaters (Gregory, 1990).

Fats

Accepted dietary recommendations are that most adults in the UK should reduce their total dietary fat intake. The 1991 UK

Dietary Reference Values; (DRV's) recommend that total fat intake for the population should be no more than 33 per cent of dietary energy (including alcohol) or 35 per cent of energy derived from food. Saturated fatty acids should provide no more than 11 per cent of food energy. The Government policy document The Health of the Nation (Department of Health, 1992) endorsed these targets.

The 1992 National Food Survey (MAFF, 1993) found that fat supplied over 41 per cent of food energy, with over 16 per cent coming from saturated fatty acids.

In general, vegetarians have a lower intake of fat than non-vegetarians (Taber and Cook, 1980). However, the percentage of food energy supplied by fats may still be higher than is desirable. Draper (1993) found 37 per cent of food energy came from fat among vegetarians. This was not significantly different from the 38 per cent among meat-eaters (Gregory, 1990). Vegan diets came closer to the ideal, with 34 per cent of food energy coming from fat. Vegetarians must ensure that they do not simply substitute high-fat dairy products or convenience vegetarian foods for meat and meat products.

While total fat intake may not always differ significantly between vegetarians and non-vegetarians, vegetarians have consistently been shown to have a lower intake of saturated fat, a higher intake of polyunsaturated fat, and a higher polyunsaturated to saturated fat (P:S) ratio. This may play a part in vegetarians' lower incidence of cardiovascular disease.

Essential Fatty Acids

Two polyunsaturated fatty acids must be present in the diet as the body is unable to synthesise them. These essential fatty acids are linoleic acid and alpha-linolenic acid. Linoleic acid is present at high levels in many vegetables, fruits, nuts, cereals and plant oils; alpha-linolenic acid is less abundant but is present in linseed, soyabean and rapeseed oils at high levels, as well as in pulses, walnuts, broccoli and leafy green vegetables.

Alpha-linolenic acid is converted to eicosapentaenoic acid (EPA) in the body, which is in turn converted to docosahexanoic acid (DHA). DHA is required for retinal and brain function and is important for foetal and infant development.

Fish Oils: Habitual consumption of fish oils have been shown to be associated with reduced mortality from heart disease. This is believed to be due to the high levels of eicosapentaenoic acid (EPA), an omega-3 fatty acid, present in fish oils.

EPA intake is low or lacking altogether on vegetarian or vegan diets. However, plant foods contain alpha-linolenic acid (ALA) which can be converted in the body to EPA. Research has shown that increasing ALA intake can lead to similar cardiovascular benefits as seen with increased EPA intake. These include reduced platelet aggregation and so a reduced risk of thrombosis (Renaud, 1986).

Vegan diets are devoid of EPA and DHA while vegetarian diets provide small amounts in dairy products. Vegetarian and vegan women should ensure their diets include sources of alpha-linolenic acid during pregnancy and breast feeding. Sanders (1992) has suggested vegans should use plant oils with a low ratio of linoleic acid to alpha-linolenic acid, eg soya bean or rapeseed oil, as a high ratio may inhibit conversion of alpha-linolenic acid to DHA. It is important that the diet has the correct balance of linoleic acid to alpha-linolenic acid and this can be achieved on a vegetarian diet.

Carbohydrates and Dietary Fibre

Dietary advice is to increase consumption of complex carbohydrates and dietary fibre (non-starch polysaccharides, or NSP) by increasing the consumption of plant foods. Compared with non-vegetarians, vegetarians typically have diets higher in both complex carbohydrate and fibre. Fibre consumption by vegetarians invariably exceed the DRV of 1.8g/day for adults, tending to range from 33g/day upwards. There has been some concern over the possibility that diets excessively high in fibre may reduce the bioavailability of certain minerals, including calcium, iron and zinc. However, populations habitually consuming high fibre diets are not characterised by mineral deficiencies.

Protein

Protein intakes of vegetarians are generally lower than those of non-vegetarians but still meet UK dietary recommendations. Jackson (1993) calculated the average intake of protein by vegetarians in the UK to be 57g/day, compared with 75g/day for non-vegetarians. Draper (1993) found protein intake in vegetarians to be 66g/day for men and 56g/day for women.

UK recommendations are 55g/day for men and 45g/day for women. Vegetarians eating a balanced diet obtain protein from a variety of food sources. This means essential amino acid requirements can be met by plant foods.

The idea of combining proteins from different plant sources to provide all of the necessary essential amino acids is known as protein complementing. Previous emphasis on protein complementing has been overstated and it is now widely accepted that it is not something vegetarians eating a balanced diet need consider.

In fact, the lower protein intake of vegetarians may be beneficial as there is concern that high protein intake may be associated with health risks. Poor or failing kidney function may be aggravated and the loss of calcium from bones increased by excessive dietary protein.

In both instances, animal protein appears to be more harmful than plant protein. A high intake of animal protein may also play a contributory role in raised blood pressure. A lower animal protein intake almost inevitably results in a lower fat intake, especially saturated fat, as foods high in animal protein also tend to be high in fat. Dietary advice is to increase the proportion of protein from plant sources (NACNE, 1983).

Vitamins

A balanced vegetarian diet supplies all the vitamins the body requires. Surveys have shown the vitamin status of vegetarian populations to be satisfactory (Shultz, 1983; Helman, 1987). Draper (1993) found intakes of most vitamins for adult vegetarians in the UK to be adequate and in many cases higher than the national average. Certain vitamins, such as vitamin B12 and vitamin D, are primarily found in animal foods and so concerns over the vitamin status of vegetarians and vegans tend to focus on these.

The only reliable unfortified sources of vitamin B12 are animal foods. Research into possible plant food sources of B12 have found that either no significant amount of B12 is present or that any B12 present is a structurally similar analogue that cannot be utilised by the body. Vegetarians generally obtain adequate B12 from dairy products. Draper (1993) found mean intake of vitamin B12 in vegetarians to be 181 per cent of the RNI in men and 121 per cent in women.

Vegans tend to have lower vitamin B12 intakes which may or may not reach recommended levels. Vegans should be advised to ensure their diet includes foods that are fortified with vitamin B12. A range of vitamin B12 fortified foods is available, including yeast extracts, vegetable stocks, vegetable burger mixes, soya milks, margarines, breakfast cereals and herbal soft drinks.

Vitamin D is present in vegetarian diets in dairy products. Vegans tend to have low vitamin D intakes, fortified margarines being the main dietary source. However, it is assumed that sufficient amounts of the vitamin can be obtained by dermal synthesis due to the action of sunlight on exposed skin.

Vegans or vegetarians consuming limited dairy products and who are not exposed to sunlight may be advised to take supplementary vitamin D. Groups at risk of insufficient vitamin D include infants and elderly persons confined indoors.

Riboflavin (vitamin B2) may also be low on vegan diets, the main dietary source being milk and milk products. Vegans should ensure their diet includes other sources of riboflavin. such as fortified breakfast cereals, yeast extract and mushrooms.

Minerals

Vegetarians have adequate mineral intakes and a balanced vegetarian diet does not lead to mineral deficiencies (Abdulla, 1984; Freeland-Graves, 1988). Studies of the mineral status of vegetarians have concentrated on iron and zinc, meat and meat products being a major source for both of these minerals. Vegetarians appear to be at no greater risk of iron deficiency than non-vegetarians and have an adequate intake of iron from plant food sources (Anderson, 1981; Levin, 1986b). Praper (1993) showed mean iron intakes of both vegetarians and vegans met the UK DRVs and indeed exceeded the mean intake of comparative non-vegetarian adults from a nationwide survey.

IRON AND DISEASE

Recent studies have suggested that high levels of iron in the body may be linked to both heart disease and cancer. Satonen (1992) studied nearly 2,000 men aged between 42 and 60 and found that both high blood ferritin levels and high dietary iron intake were linked to heart attack risk. Similarly, Stevens (1994) found that people with elevated levels of iron in the body developed cancer at nearly twice the rate of those with normal iron levels.

It is suggested iron can cause cell and tissue damage in the body by generating free radicals. As haem iron in meat is more readily absorbed than non-haem iron on plants then excess meat consumption may be linked with chronic disease.

Non-haem iron from plant foods is less readily absorbed by the body than haem iron from meat and meat products. However, vitamin C greatly improves non-haem iron absorption and vegetarian diets are generally rich in vitamin C.

It is recommended that plant food sources of iron are consumed at the same time as vitamin C rich foods or drinks to facilitate absorption. Good sources of iron for vegetarians include leafy green vegetables, pulses (including baked beans), dried fruit, wholemeal bread and fortified breakfast cereals. There is some suggestion that vegetarians may adapt to their diet by an increased ability to absorb iron (Anderson, 1981). For this reason, new vegetarians may be at a greater risk of iron deficiency than long-term vegetarians (Helman, 1987).

Zinc status among vegetarians also appears to be adequate (Levin, 1986b; Draper, 1993). Freeland-Graves (1980) found no significant differences in mean zinc intakes between vegetarians and non-vegetarians. However, some vegetarian women had low zinc intakes and this was attributed to limited consumption of zinc-rich foods such as pulses and wholegrains. Low zinc intake reflects poor food choices rather than inadequate food sources.

The bioavailability of both iron and zinc is reduced by fibre, phytates and other dietary factors present in plant foods. However, Anderson (1981) found the iron and zinc status of vegetarian women was normal despite high fibre and phytate intake while other studies have confirmed that a high fibre intake does not appear to affect mineral utilisation by vegetarians (Rattan, 1986; Kelsay 1988).

Due to their avoidance of dairy products it has been suggested that vegans may be at risk of calcium deficiency. However, studies have indicated that vegan populations receive adequate calcium from plant foods (Abdulla, 1981, Draper, 1993). Vegetarians have calcium intakes similar to or exceeding those of non-vegetarians.

There is also evidence to suggest vegans may be at risk of iodine deficiency, milk and milk products being the main dietary source (Key, 1992; Draper, 1993). However, many vegans consume edible seaweeds which are good sources. Cereals and vegetables can also contain available amounts, depending on iodine levels in the soil in which they are grown.

SPECIAL DIETARY GROUPS

Some groups may have special dietary needs. These include pregnant and lactating women, infants and children, and those following a therapeutic diet for medical reasons. The needs of all these groups can be adequately met by a vegetarian diet.

Pregnancy and Lactation

During pregnancy, women have increased nutritional requirements for energy, protein, folate, vitamin A, vitamin C, and vitamin D. The increased energy and protein requirements are small and can be met without any obvious changes in food intake. If energy intake is reduced due to appetite changes or morning sickness then a reduction in high fibre foods and increased energy and nutrient dense foods may be appropriate. The extra vitamins needed are generally present at higher levels in vegetarian than non-vegetarian diets, with the exception of vitamin D. Both vegetarian and non-vegetarian pregnant women may be advised to take supplementary vitamin D.

There are no official dietary recommendations for increased mineral requirements during pregnancy. However, it is important that adequate intakes are ensured in order to maintain body stores.

Ideally, all women should have sufficient body iron stores to cope with the demands of pregnancy. However, many women may have low levels of stored iron and mild anaemia is not uncommon during pregnancy. Vegetarian women need to ensure that they are obtaining adequate iron from their diets.

Vegan women need to ensure that their dietary needs for vitamin B12, vitamin D and calcium are met during pregnancy. During lactation, increased protein and vitamin needs can be met by a varied vegetarian diet. Recommended intakes for calcium and zinc are also increased and so extra food sources of these may be appropriate.

Infants and Children

A vegetarian diet is suitable for children of all ages and can provide all the nutrients needed for normal growth and development.

Studies have found that vegetarian children show similar patterns of growth and development as non-vegetarian children. Anthropometric measurements indicate vegetarian children are similar in height, weight and skinfold measurements.

Vegetarian children are often leaner and less likely to be obese. Sabate (1991) studied 895 non-vegetarian children and 870 vegetarian children aged 7 to 18 years and found that for both boys and girls the vegetarian children were taller. This did not change after adjustment for parental height and socioeconomic class. The vegetarian children consumed less convenience foods and dairy products and more starchy foods, pulses, fruit and vegetables which may play a part in their being taller. No differences in maturation or timing of growth spurts has been observed between vegetarian and non-vegetarian boys. Vegetarian girls may have a delayed maturation onset, though heights finally attained are similar to non-vegetarian counterparts (Sabate, 1992). This maturation delay is linked with a later age of menarche which may be protective against breast and other hormone dependant cancers in later life.

Children raised on vegan diets also display growth and development within acceptable limits though they are more likely to be lighter and leaner and may also be shorter in stature than other children.

Where concern has been expressed over the growth of infants and children on vegetarian-type diets it has involved communities where diets followed are very restrictive and not typical vegetarian or vegan diets. These have included Dutch children on macrobiotic diets and Israeli Black Hebrews.

Infants, children and adolescents can meet all of their nutritional requirements on a vegetarian diet. Nutrient intakes for vegetarian children compare favourably with dietary recommendations providing that total fat intakes are not excessive and iron intakes are adequate.

Nutritional deficiencies are no more commonly reported among vegetarian children than among omnivore children. Vegan children may need more carefully planned diets to ensure nutrient needs are met, especially for vitamin B12, vitamin D and calcium.

As well as meeting dietary needs, vegetarian diets for children may be beneficial in promoting health and protecting against a variety of health problems including constipation, other bowel disorders and obesity The typical diet of school children in the UK is high in fat and low in fibre and complex carbohydrates and it is recommended children eat more starchy foods, fruit and vegetables.

A vegetarian diet can also protect against risk factors for chronic degenerative diseases in later life (Jacobs, 1988). Risk factors for diet-related diseases, including cardiovascular disease, can be identified in children and dietary habits which influence these risks are learnt in childhood. A vegetarian diet for children can help establish healthy dietary patterns which may be carried on into adult life and help protect against chronic disease.

A healthy adult diet is not necessarily appropriate for infants and young children under the age of five years. Diets low in fat and high in fibre are unsuitable for infants as their limited stomach capacity may be filled-up before they are able to meet their energy or nutrient needs. Infant diets need to include energy and nutrient dense foods such as cereal products, mashed lentils, vegetable oils, bananas and avocados. Excessive intake of high fibre or watery foods should be avoided.

Therapeutic Diets

As well as being protective against many chronic degenerative diseases, vegetarian diets have been effectively used in the treatment of various conditions including rheumatoid arthritis (Kjeldsen-Kragh, 1991) and nephrotic syndrome (D'Amico, 1992). A vegetarian diet is also appropriate for most people following a therapeutic diet for medical conditions, such as diabetes or coeliac disease. Professional dietetic assistance should be sought for all therapeutic diets.

Conclusion

1. A vegetarian diet can confer a wide range of health benefits and may protect against chronic degenerative diseases including coronary heart disease and diet related cancer.
2. Blood cholesterol level is typically 10 per cent lower in vegetarian groups compared with non-vegetarian populations. A 10 per cent lowering of blood cholesterol may result in up to a 50 per cent reduction in the incidence of heart disease.
3. A vegetarian diet is more effective at lowering blood cholesterol and reducing the risk of cardiovascular disease than a diet in which lean meat is regularly consumed. A low-fat vegetarian diet has been shown to be acceptable

to heart patients and high levels of dietary compliance have been demonstrated.

4. A balanced vegetarian diet tends to correspond closely with current dietary advice for healthy eating. Vegetarians tend to have lower intakes of total and saturated fat and higher intakes of complex carbohydrates and dietary fibre.
5. Despite vegetarian diets generally being lower in fat, the percentage of food energy from fat may still be higher than dietary recommendations. Vegetarians need to ensure that consumption of high-fat foods, especially dairy products, convenience foods and margarine and plant oils, is not excessive.
6. A balanced vegetarian diet can supply all the vitamins and minerals required. Vitamin and mineral status in vegetarian groups is satisfactory and intakes are often higher than in non-vegetarians. Vegetarian diets high in fibre do not appear to compromise mineral status.
7. Vegetarians generally have satisfactory iron levels. It is recommended that dietary sources of iron are consumed together with vitamin C rich foods or drinks to facilitate iron absorption.
8. Vegans should ensure their diets include foods which are fortified with vitamin B12. These include certain yeast extracts, vegetable burger mixes, soya milks, margarines and breakfast cereals.
9. A vegetarian diet is appropriate for infants and children and provides all the nutrients required for normal growth and development. Vegetarian children are similar in height and weight to non-vegetarian children and are less likely to be obese or overweight.
10. A healthy vegetarian diet for adults is not appropriate for infants or children below the age of 5 years. Diets low in fat and high in fibre can fill up infants before they have received adequate energy and nutrients. Infant diets need to include fewer fibre rich foods and more energy and nutrient dense foods.

DIET AND DISEASES

CANCER

Cancer is the second leading cause of death in Britain, accounting for 25% of all deaths. It has been estimated that diet may be linked to 30-70% of cancers (Doll, 1990). Certain cancers, such as colon, breast and prostate are clearly diet related (Cummings & Bingham, 1998).

Sir Kenneth Calman, Chief Medical Officer, has stated (1997) that "there is a relationship between eating red meat and cancer".

The Oxford Vegetarian Study found cancer mortality to be 39% lower among vegetarians compared with meat-eaters (Thorogood, 1994).

A study of 23,000 largely vegetarian Seventh Day Adventists found cancer mortality rates to be 50-70% of those of the general population for several cancer sites unrelated to smoking or alcohol (Phillips, 1975).

Professor Nick Day of the University of Cambridge and the European Prospective Study into Cancer has stated that vegetarians may suffer 40% fewer cancers than the general population.

The World Cancer Research Fund's dietary advice to minimise cancer risk involve reducing the intake of dietary fat and increasing the consumption of fruits, vegetables and wholegrains.

COLON CANCER

Vegetarians have lower rates of colon cancer than non-vegetarians (Phillips, 1980). Incidence of colon cancer has been strongly linked to the consumption of meat (Armstrong, 1975, Singh & fraser, 1998). Willett (1990) carried out a study of over 88 000 women aged 34 to 59 years. Women eating red meat daily ran over twice the risk of developing colon cancer than women eating red meat less than once a month.

Reduced incidence of colon cancer in vegetarians may be attributed to dietary differences which include increased fibre intake, increased consumption of fruit and vegetables, and decreased intake of total fat and saturated fat. The mechanism by which a vegetarian diet is protective against colon cancer is unclear and a great deal of research is being carried out in this area.

It has been suggested that secondary bile acids are carcinogens which may play an important role in colon cancer. These are derived by bacterial metabolism from primary bile acids made in the liver and secreted into the intestine. Vegetarians have lower levels of secondary bile acids than non-vegetarians (Turjiman, 1984). The differences in bacterial populations between the intestines of vegetarians and non-vegetarians may also be important. Bacterial flora in vegetarians has been shown to possess reduced ability to transform bile acids into potential carcinogens (Johansson, 1990).

The role of dietary fibre in prevention of colon cancer may also be important. This was first noted in 1971 when it was suggested the high incidence of colon cancer in Western countries was linked to low fibre diets. Other dietary components associated with high fibre foods, such as folate, have also been implicated as having protective effects.

Chen (2002) found that the risk of distal stomach adenocarcinoma was positively associated with red meat intake.

BREAST CANCER

Evidence also suggests a vegetarian diet is protective against breast cancer (Phillips, 1975). This may be due to the increased fibre and reduced fat intake of vegetarian diets. Vegetarian diets can alter the levels of circulating sex hormones which may have a beneficial effect. Fibre is thought to be protective by modifying circulating oestrogen levels.

Studies of adolescent girls have shown age of menarche to be delayed in vegetarians (Sabate, 1992). Later age of menarche is believed to lower the risk of breast cancer in adult life.

OTHER CANCERS

Studies have shown vegetarians to suffer less from various other cancers.

Mills (1989) studied the incidence of prostate cancer amongst 14,000 Seventh Day Adventists and found a relationship between increased risk and increasing animal product consumption.

Mills (1988) also found pancreatic cancer to be associated with consumption of animal products. Increasing consumption of fruit, vegetables and pulses was shown to have a protective effect.

Rao (1989) found a vegetarian diet to be protective against oesophageal cancer.

Studies have also shown vegetarians to have lower incidence of lung cancer. This can be largely attributed to vegetarians tending to be non-smokers. High consumption of fruit has also shown to be protective against lung cancer (Fraser, 1991).

DIVERTICULAR DISEASE

Diverticular disease affects the colon and symptoms include lower abdominal pain and disturbed bowel habit. It occurs frequently in western countries where intake of dietary fibre is low. Gear (1979) found diverticular disease to be less frequent in vegetarians, 12% of vegetarians studied having diverticular disease compared with 33% of non-vegetarians. This is thought to be due to the increased fibre of vegetarian diets.

Gall Stones

Gall stones are composed of cholesterol, bile pigments and calcium salts. They form in the gall bladder and can cause severe pain. A study of over 750 women found the incidence of gall stones to be less frequent in vegetarians. 25% of non-vegetarians compared with 12% of vegetarians had gall stones. After controlling for age and body weight, non-vegetarians were found to have a relative risk of gall stones almost twice that of the vegetarians. Vegetarians are leaner, and consume more dietary fibre and less dietary cholesterol, all of which is believed to protect against gall stone formation.

Kidney Stones

Kidney stones form in the kidney and can cause considerable pain when passing down the urinary tract. Prevalence of kidney stones is lower in vegetarians. A high intake of animal protein increases the urinary loss of calcium and oxalate, known risk factors in kidney stone formation. Meat is also high in purines which leads to increased uric acid in the urine. Urinary uric acid is also a risk factor for kidney stones.

Osteoporosis

Osteoporosis is the loss of calcium from bone tissue, leading to bones that are brittle and liable to fracture. It is most commonly seen in postmenopausal women.

Some studies have suggested that vegetarians may be at lower risk of osteoporosis than non-vegetarians. Sellmeyer's study (2001) found that elderly women with a high dietary ratio of animal to vegetable protein intake had more rapid femoral neck bone loss and a greater risk of hip fracture than those with a low ratio.

Marsh (1988) found bone loss to be considerably less in postmenopausal women who were vegetarian than those who were non-vegetarian. The non-vegetarian diet contained higher amounts of sulphur, which derived from animal protein. Dietary sulphur increases the acidity of urine, which results in increased urinary calcium loss. Increased urinary calcium loss is related to increased calcium loss from bone tissue.

Hip fractures associated with osteoporosis has been shown to be higher in countries consuming a diet high in animal protein.

Appendicitis

The Oxford Vegetarian Study found that people who do not eat meat have a 50% lower risk of requiring an emergency appendicectomy that those who do.

OTHER DISEASES

A vegetarian diet has been claimed to reduce the risk of gout, hiatus hernia, constipation, haemorrhoids, and varicose veins. These diseases are linked to diets low in fibre and high in saturated fat.

Food Poisoning and Pesticide Residues

Over 50,000 cases of food poisoning are reported every year and the actual incidence of food poisoning is estimated to be ten times this figure. Meat, eggs and dairy products are the primary sources of food poisoning. Professor Richard Lacey of the University of Leeds has stated that "More than 95% of food poisoning is derived from meat and poultry products".

Pesticide residues in foods include PCB's and dioxins. These are found in highest concentrations in meat, fish and dairy products.

Studies have shown these toxic chemicals can be passed on from pregnant women to infants during both pregnancy and lactation and may damage the developing nervous systems. Hall (1992) has stated a vegetarian diet minimises the risk of contamination.

Rheumatoid Arthritis

Studies have shown that vegetarian diets can be successfully used to treat the symptoms of rheumatoid arthritis and other rheumatic diseases. Kjeldsen-Kragh (1991) found that rheumatoid arthritis patients following a vegetarian diet suffered considerably fewer swollen and tender joints and less stiffness or pain.

Nephrotic Syndrome

Nephrotic syndrome is a kidney condition involving high levels of protein in the urine which may lead to progressive kidney damage as well as promoting atherosclerosis and heart disease. Studies have shown a low protein vegan diet can be used to reduce the symptoms of nephrotic syndrome (D'Amico, 1992).

The China Health Project

The China Project on Nutrition, Health & Environment is a massive study involving researchers from China, Cornell University in Ithaca, New York, and the University of Oxford, into the relationships between diet, lifestyles and disease-related mortality in 6500 Chinese subjects from 65 mostly rural or semi-rural counties.

The rural Chinese diet is largely vegetarian or vegan, and involves less total protein, less animal protein, less total fat and animal fat, and more carbohydrate and fibre than the average Western diet. Blood cholesterol levels are significantly lower.

Heart disease, cancer, obesity, diabetes, and osteoporosis are all uncommon. Areas in which they are becoming more frequent are areas where the population has moved towards a more Western diet with increasing consumption of animal products.

The China Health Project has clearly demonstrated the health benefits of a diet based on plant foods. One of the Project's co-ordinators, Dr Colin Campbell of Cornell University, has stated that "We're basically a vegetarian species and should be eating a wide variety of plant foods and minimising our intake of animal foods."

INCREASE FOOD INTAKE

If you have lost a lot of weight due to illness, your doctor or dietitian may advise you to increase your intake of energy and protein. They may try to persuade you that your vegetarian diet is unsuitable. Do stick to your principles. This Information Sheet is to enable you to choose vegetarian foods which will help you to gain weight and recover from illness.

The energy density of foods is important when trying to gain weight. Energy density refers to the amount of energy or calories compared to the weight of the food. Vegetable foods tend to be less energy dense and so it is particularly important for vegetarians to be aware of suitable foods.

When you are poorly, modern healthy eating advice to reduce fat, reduce sugar and increase fibre, is peripheral to your immediate needs. If you know you must increase your energy and protein intake, you can do so by increasing your fat and sugar intake, and also try to avoid large amounts of fibre.

If You are Underweight but Healthy

If you are not ill but feel you are underweight and need to gain then not all the advice on this sheet is suitable for you. You should try and increase your food intake without consuming too much fat and sugar although a little with the right type of food is fine. For example, you can use olive oil on baked or boiled potatoes and sugar in a nutritious wholemeal fruit cake.

Regular meals, especially breakfast, are important if you find it difficult to gain weight. Taking your time with meals and making sure you are relaxed is also important to ensure that your digestive system functions properly. A glass of wine with an evening meal is beneficial to relax you as well as stimulating your digestive juices.

Some people are just naturally thin no matter how much they eat and the envy of those who gain weight easily. As long as you are in good health and do not drop below the recommended range of weights for your height you should not worry. Exercise can help build up your muscles and make you feel better about your body shape.

Lack of Appetite

If your appetite is poor, you must not forget to eat. Try to have small but frequent meals, and if you can manage, snacks in between. You may like to try nutritious drinks between meals.

If you feel nauseous, try to avoid strong cooking smells and have a short walk or perhaps just sit in the fresh air before a meal.

PROTEIN

Protein is essential to the body, for repair, to enable the immune system to function and for recovery from illness. If you are recovering from illness, it is important to include adequate protein in your diet. Good vegetarian sources of protein include: cereals, nuts & seeds, soya products (soya milk, tofu, tempeh etc) and pulses. Milk, yoghurt, cheese, and free range eggs are also excellent protein sources for the non-vegan.

You must ensure that you include a protein source at each meal. Try to choose the energy dense sources of protein that are relatively soft and which do not contain to much fibre, for example, the soya or dairy sources of protein.

Try nut butters. There are many more available other than peanut butter, try others such as hazel nut or cashew nut butter.

Soya products are excellent and easily digested sources of protein. Choose the nutritional supplemented soya milks with added calcium and sugar.

If you use dairy products, make good use of soft milky puddings, such as custard and rice pudding. Vegans can also get a variety of soya pudding, including soya rice pudding. If you like, add jam, honey, ground nuts or cream to your pudding. Porridge with a couple of tablespoons of added vegetable oil, perhaps with some nuts and/or dried fruit, can be very nutritious. If you prefer to have white bread, white pasta and refined rice, then do so. But remember that the wholemeal varieties are very good sources of the B vitamins.

Make good use of cereal products (wheat, barley, rye, oats etc.), although avoid too much fibre. Complement cereals with other sources of proteins, such as nuts, seeds, pulses, milk and cheese.

Seeds, such as sesame seeds (tahini) or sunflower seeds, when ground as a paste, can also provide an energy dense source of protein in the diet. Tahini is particularly good when combined with a mashed pulse, such as chick peas to make hummus.

Pulses are more easily digested, if well cooked and mashed. Try mashed lentils or pureed black eye beans.

ENERGY

In order to gain weight it is essential to increase the energy content of your diet. Also, in order for your body to make proper use of protein, you must ensure sufficient energy. The most concentrated form of energy in the diet is from fat. If you can tolerate fat then try and increase the fat content of your diet. Alcohol is also a concentrated form of energy and can be taken in moderation, perhaps before meals, as it may help improve your appetite. Only take alcohol with your doctors permission.

If you can tolerate oils, try adding some vegetable oil or vegetable margarine to your food, particularly to mashed pulses, salads and potatoes. If you like fried foods, then have fried foods. Use cream if you can tolerate fats. A good vegan alternative to cream is cashew nut cream, made from ground cashew nuts and water.

Do not restrict your sugar intake, although always have sweet foods after the savoury part of your meal. Glucose is a less sweet and more concentrated form of sugar. Glucose can be added to foods and drinks, such as porridge and cocoa drinks to increase your energy intake. Check with your dietitian about using glucose. She may suggest that your doctor prescribes you a special type of non-sweet, high calorie supplement, which can be added to foods and drinks.

If you like to have soup, make sure the soup is as nutritious as possible. Add pureed lentils, vegetable oil, butter, whole milk, cream or cheese.

Make good use of the wide range of vegetable pates and spreads available.

GENERAL ADVICE

- You may prefer to take puddings half an hour to one hour after your first course, this will enable time for the first course to settle.
- You may find ice-cream, soya or dairy, refreshing after meals.
- Make good use of a food blender, if you have one, to blend fruit, vegetable and pulses.
- Have meals with a good sauce, made with pureed fried onions, perhaps with tomatoes or mushrooms. To increase the energy density of sauces use butter or vegan margarine and add soya milk, silken tofu, cow's milk or cream.
- Do make sure you include a source of vitamin C in your diet, perhaps fresh fruit juice, black-currant cordial or rosehip syrup.
- Try and eat some fresh fruit and vegetables or salad each day, especially green vegetables, to ensure you get your

full compliment of vitamins and minerals. If you cannot tolerate whole fruit or crunchy vegetables, take them as fruit juice or a pureed soup made with fresh vegetables.

- High energy drinks, can be made using whole cow's milk, sweetened soya milk or silken tofu, with glucose, honey or sugar and a flavouring of your choice such as chocolate or vanilla.
- You can buy products from the chemists, to have as meal replacements, when your appetite is particularly poor. Do check on their suitability for vegetarians.
- Ask your doctor or dietitian about including a vitamin or mineral supplement in your diet. If you choose to take a supplement, inform your doctor and dietitian. Multi-vitamin and mineral tablets should not do you any harm, as long as you stick to the recommended dose.

If You Still Need Extra Protein & Energy

Your doctor can prescribe for you special supplements which can help you to gain weight. You would usually be referred to a dietitian, who can nutritional assess you and recommend the correct type of supplement. If you are concerned about the suitability of a any product, for vegetarians or vegans, which have been prescribed for you, contact your doctor. Also contact us if you are having problems with health professionals who may be sceptical about your vegetarian or vegan diet.

BALANCE OF GOOD HEALTH

A "portion" is, for example, a slice of bread, an apple, a glass of milk or two tablespoons of baked beans, but remember-it's the balance that matters so if you have a big appetite, increase the amount you eat in all categories, not just the fatty and sugary foods.

Fruit and Vegetables

5 portions a day: Fresh, frozen, juiced, tinned or dried fruit and vegetables are particularly good sources of vitamins, minerals and fibre.

Bread, other Cereals and Potatoes

5 portions a day: Base most of your meals on these starchy foods.

Use wholemeal or wholegrain versions as much as possible and avoid adding lots of fat. These foods give us carbohydrates for energy, fibre, protein and some vitamins and minerals.

High Protein Foods

2-3 portions a day: Include a variety of pulses (for example beans, lentils, peas and chick peas), nuts, seeds, eggs, soya, mycoprotein (Quorn) or wheat proteins to give you plenty of protein, minerals and vitamins.

Milk and Dairy Products

2-3 portions a day: Good sources of calcium, protein and some vitamins. If you are avoiding dairy foods, choose fortified soya, rice or oat drinks or make sure that you eat other foods that are high in calcium.

Fatty and Sugary Foods

0-3 portions a day: Although we need to eat some fat, we all need to eat these foods sparingly and try low fat alternatives.

33% of the UK population only eat meat occasionally.

Gallup Poll for Realeat, August 2001

5% of the UK population is vegetarian

National Diet and Nutrition Survey 2001

"The British Dietetic Association considers that a well-planned vegetarian diet can be nutritionally adequate and health promoting for both adults and children."

Particular Nutrients

Although a balanced vegetarian diet is lower in fat and higher in fibre and anti-oxidants than a meat-based diet, many people still believe that vegetarians can't possibly be healthy because they think certain nutrients are only found in meat. This is simply not true.

Iron

Iron deficiency is one of the most common nutritional problems in a typical British diet. Research shows that vegetarians are no more prone to iron deficiency than meat-eaters and even those who do eat meat get 86% of their iron from vegetarian sources. Iron is found in leafy green vegetables, pulses, wholemeal bread, dried fruit and pumpkin seeds. If you are worried about getting enough iron, try eating something high in Vitamin C (or drinking a glass of orange juice) with a meal containing iron-rich foods because this can triple the amount of iron your body is able to absorb.

Protein

We need to eat eight different types of amino acid-the building blocks for protein. Most vegetarian sources don't include all eight, so are sometimes thought of as 'incomplete' protein, but a varied vegetarian diet does include them all-just from different sources. Your body keeps a pool of the amino acids it needs, so all you have to do is eat a variety of different high-protein foods through the week.

Good vegetarian sources of protein include pulses (such as beans, lentils, peas or chick peas), nuts, seeds, grains, soya (eg tofu, soya mince), dairy products and eggs.

Vitamin B12

B12 is not naturally found in plant foods, so vegans need to eat fortified foods such as breakfast cereals and soya drinks, but anyone eating dairy products and eggs will get plenty of B12 from those foods.

Fatty Acids

The omega 3 essential fatty acids found in oily fish are also found in vegetarian foods such as rapeseed oil, flax seeds and walnuts.

So, what do you eat?

Eating vegetarian is a real culinary adventure. Once you've broken out of the conventional meat-and-two-veg, you can break all the rules, or stick with more familiar tastes if you prefer. Many of your favourite dishes, such as spaghetti bolognaise, chilli, stir-fry and curries can be made with soya mince, pulses or just vegetables. Bean burgers, pizzas, veggie sausages, ratatouille, baked beans, scrambled eggs, jacket potatoes, many cheeses and all sorts of ready meals, pasta sauces and soups are all suitable for vegetarians already.

A few recipe ideas are shown on the following pages, but before you start cooking, fill your cupboards with the right basic ingredients and you'll find the transition much easier, and tastier.

THE VEGETARIAN STORE CUPBOARD

Most people keep a few stock items in their kitchen cupboard or freezer. Many of the basics-bread, flour, vegetable oil, dairy products, free-range eggs, frozen chips etc-will be suitable for vegetarians, but you may find it helpful to use the following as a guide for your first big shop as a new veggie.

- Lots of fresh fruit and vegetables
- Vegetable oil
- Olive oil
- Vegetable stock cubes
- Vegetarian gravy granules (check the label, many meat-flavoured varieties are actually vegetarian)
- Yeast extract (eg marmite)
- White wine vinegar (or balsamic for a treat)
- Peanut butter
- Tinned beans and pulses
- Tinned tomatoes
- Tinned soup
- Dried pasta
- Rice
- Quick-cook noodles
- Cous-cous

- o Dried soya chunks
- o Tinned soups
- o Ready made pasta/curry sauces
- o Dried herbs and spices
- o Seeds (try sesame, pumpkin, sunflower and many others)

To Keep in the Fridge

- o Free range eggs
- o Milk
- o Vegetarian cheese
- o Butter or margarine
- o Natural yoghurt
- o Houmous
- o Vegetarian pesto
- o Tofu (look out for marinated or smoked varieties)
- o Tomato puree
- o Mustard
- o Quorn products (look for the Vegetarian Society approved products)
- o Jars of sun-dried tomatoes or other antipasto (mushrooms, olives, artichokes or peppers in seasoned oil)
- o Fresh soups, pies and ready-meals

To Keep in the Freezer

- o Soya and Quorn mince or chunks
- o Vegetarian sausages and burgers
- o Ready rolled pastry
- o Pizza bases
- o Tortilla wraps
- o A variety of frozen vegetables
- o Ready-made pies and meals (many supermarket freezers have a special vegetarian section)

o Shelled nuts (try brazils, cashews, almonds and walnuts)-buy in bulk to save on the price and store in the freezer to keep them tasting fresh.

RECIPES AND MEAL IDEAS

Roast Mushroom and Aubergine Wrap

Place mushrooms, pieces of aubergine, courgette and tomato in a baking tray and drizzle with olive oil. Bake until tender, at 200 C/400 F/Gas Mark 6. Once cooked, leave to cool, add a few pitted olives and wrap with a flour tortilla or use to fill a pitta.

The Hallelujah Baguette!

Fry a few slices of halloumi on a high heat until golden brown. Pop into a baguette, along with some chopped sun-dried tomatoes, fresh salad leaves and a few springs of fresh mint.

Mushroom Coronation Ciabatta

Gently fry a punnet of oyster mushrooms in olive oil with 4 or 5 spring onions (sliced) and 1tbsp of curry powder. Stir in1 tsp of sweet chutney or apricot jam, cool then add 2 tbsp of mayonnaise. Chill the mixture and use to fill a ciabatta or soft white rolls.

The RLT

Veggies don't have to miss out on old favourites-grill or fry some vegetarian rashers (or make your own by frying thinly sliced tempeh with soy sauce), sandwich between sliced bread on a bed of crisp lettuce and sliced tomato, add veggie mayo or relish to taste. Grated Carrot and Sultana Sandwich with Peanut Butter. Lightly toast 2 slices of granary bread and allow to cool. Spread one slice with margarine and one with peanut butter. Grate a carrot, mix with a handful of sultanas and pile onto the peanut butter. Add some shredded iceberg lettuce and finely sliced onion if you wish. Sandwich and scoff!

Avocado and Houmous Bagel

Fill your bagel with lashings of houmous and a layer of mashed or sliced avocado. Top with slices of tomato and cucumber and a good dose of coarse ground black pepper.

Pitfalls

Once you get used to cutting meat out of your diet you can start checking labels for non-vegetarian ingredients, which include lots of by-products of the slaughter house as well as obvious things like meat, poultry and fish. Most vegetarians also choose to avoid battery eggs.

Luckily many vegetarian products are clearly marked as being suitable, including thousands of products which have our 'Vegetarian Society Approved' seedling symbol, guaranteeing that the product is 100% suitable to our own strict standards (including only free-range eggs).

If you come across products that aren't clearly marked, it is best to take a look at the label. Common ingredients to steer clear of are gelatine (made from the bones and skin of slaughtered cows and pigs), animal fat (which always means carcass fat), fish oils and cochineal (E120-crushed insects), which are never suitable for vegetarians. Lots of ready-meals and other products also contain battery eggs-even if they are marked as 'suitable for vegetarians'.

Unfortunately there are also quite a lot of ingredients that can be made from both vegetarian and non vegetarian sources, so if you see them listed and the product is not marked as being vegetarian, you have to assume the worst. Ingredients that fit into this category include glycerine, lactose, magnesium stearate, stearic acid and whey, together with these E numbers opposite.

E numbers marked * may be theoretically made in a non-vegetarian version, but are practically always suitable.

Don't worry-the list looks long but there are an awful lot of e-numbers in the world, so they don't turn up on as many foods as you might think!

Products To Watch Out For

- *Beer & Wine:* Some will have been fined (or made clear) with isinglass, a fish gelatine, or other animal products.
- *Cheese:* A lot of cheese produced today is entirely vegetarian, and is often marked as such, but be careful-especially with continental cheeses-because traditional rennet (used to

make the cheese set) is taken from the stomachs of young calves.

- o *Chocolate Bars and Sweets:* Watch out for gelatine, non-vegetarian whey (made with traditional rennet) and cochineal in some brands.
- o *Worcestershire Sauce:* Some brands contain anchovies.
- o *Thai Curry Sauce and Paste:* May contain fish or fish sauce.

Beyond What You Eat: Many vegetarians prefer to avoid wearing leather and there is now a bigger choice of non-leather shoes and other goods available than ever before. As well as high-street stores, you may want to try specialist shops and mail order outlets.

Toiletries and cosmetics are not all free from animal ingredients. For example, sodium tallowate-a common ingredient in soap-is an animal fat and toothpastes often contain glycerine, which may be derived from animal sources.

Our website is a great resource for seeking out vegetarian-friendly clothing, toiletries, cosmetics and all sorts of other goods, either through our online shop or links to manufacturers and retailers. Vegetarian Society Approved products don't include any animal ingredients and have not been tested on animals either.

HEALTHY EATING FOR VEGETARIANS

A varied vegetarian diet will supply all the essential nutrients you need to help you keep fit and healthy. In fact, a vegetarian diet can offer a wide range of health benefits. Research has shown that vegetarians suffer less from obesity, coronary heart disease, high blood pressure, type II diabetes, certain diet related cancers, diverticular disease, appendicitis, constipation and gallstones. A healthy diet includes plenty of fruit and vegetables and starchy foods, moderate amounts of high protein foods (which you may think of as alternatives to meat and fish), dairy produce or alternatives, and small amounts of fatty and sugary foods. You don't need to obey strict rules every day, but the following advice gives you a general idea of how to balance different types of food.

If you eat broadly the proportions described, with plenty of variety, you shouldn't have any problems achieving a healthy, balanced diet.

A BALANCED DIET

A balanced diet must contain carbohydrate, protein, fat, vitamins, mineral salts and fibre. It must contain these things in the correct proportions.

1. *Carbohydrates*: these provide a source of energy.
2. *Proteins*: these provide a source of materials for growth and repair.
3. Fats: these provide a source of energy and contain fat soluble vitamins.
4. *Vitamins*: these are required in very small quantities to keep you healthy.
5. *Mineral Salts*: these are required for healthy teeth, bones, muscles etc..
6. *Fibre*: this is required to help your intestines function correctly; it is not digested.
7. *Balanced Diets*: we must have the above items in the correct proportions.

Carbohydrates

Carbohydrates are the most important source of energy. They contain the elements Carbon, Hydrogen and Oxygen. The first part of the name "carbo-" means that they contain Carbon. The second part of the name "-hydr-" means that they contain Hydrogen. The third part of the name "-ate-" means that they contain Oxygen. In all carbohydrates the ratio of Hydrogen atoms to Oxygen atoms is 2:1 just like water.

We obtain most of our carbohydrate in the form of starch. This is found in potato, rice, spaghetti, yams, bread and cereals. Our digestive system turns all this starch into another carbohydrate called glucose. Glucose is carried around the body in the blood and is used by our tissues as a source of energy. Any glucose in our food is absorbed without the need for digestion. We also get some of our carbohydrate in the form of sucrose; this is the sugar which we put in our tea and coffee (three heaped spoonfuls for me!). Both sucrose and glucose are sugars, but sucrose molecules are too big to get into the blood, so the digestive system turns it into glucose.

When we use glucose in tissue respiration we need Oxygen. This process produces Carbon Dioxide and water and releases energy for other processes. Now try my starch test in the Virtual Laboratory

Proteins

Proteins are required for growth and repair. Proteins contain Carbon, Hydrogen, Oxygen, Nitrogen and sometimes Sulphur. Proteins are very large molecules, so they cannot get directly into our blood; they must be turned into amino-acids by the digestive system. There are over 20 different amino-acids. Our bodies can turn the amino-acids back into protein. When our cells do this they have to put the amino-acids together in the correct order. There are many millions of possible combinations or sequences of amino-acids; it is our DNA which contains the information about how to make proteins. Our cells get their amino-acids from the blood. Now try my Biuret test in the Virtual Laboratory

Proteins can also be used as a source of energy. When excess amino-acids are removed from the body the Nitrogen is excreted as a chemical called urea. The liver makes urea and the kidney puts the urea into our urine.

Fats

Like carbohydrates, fats contain the elements Carbon, Hydrogen and Oxygen. Fats are used as a source of energy: they are also stored beneath the skin helping to insulate us against the cold. Do not think that by avoiding fat in your diet you will stay thin and elegant! If you eat too much carbohydrate and protein, you will convert some of it into fat, so you will put on weight. You must balance the amount of energy containing foods with the amount of energy that you use when you take exercise. You must have some fat in your diet because it contains fat soluble vitamins.

Vitamins

Vitamins are only required in very small quantities. There is no chemical similarity between these chemicals; the similarity between them is entirely biological.

Vitamin A: good for your eyes.

Vitamin B: about 12 different chemicals.

Vitamin C: needed for your body to repair itself.

Vitamin D: can be made in your skin, needed for absorption of Calcium.

Vitamin E: the nice one-reproduction?

Mineral Salts

These are also needed in small quantities, but we need more of these than we need of vitamins.

Iron: required to make haemoglobin.

Calcium: required for healthy teeth, bones and muscles.

Sodium: all cells need this, especially nerve cells.

Iodine: used to make a hormone called thyroxin.

Fibre

We do not/can not digest cellulose. This is a carbohydrate used by plants to make their cell walls. It is also called roughage. If you do not eat foods materials which contain fibre you might end up with problems of the colon and rectum. The muscles of you digestive system mix food with the digestive juices and push food along the intestines by peristalsis; if there is no fibre in your diet these movements cannot work properly.

A Balanced Diet

You must have carbohydrate, protein, fat, vitamins, minerals salts and fibre in the correct proportions. If there is not enough protein, you will not be able to grow properly and you will not be able to repair yourself i.e. wounds will not heal properly. If you do not have enough energy containing foods you will feel very tired, you will not have enough energy. If you have too much energy containing foods you will become overweight. If you think that you are overweight you might try taking more exercise to "burn off" some of the excess food which you ate at you last meal.

FATTY ACIDS WITH TRANS CONFIGURATION

Most of the unsaturated fatty acids in nature have a cis configuration. The trans configuration results mainly during chemical hydrogenation of polyunsaturated fatty acids.

In February 1995 the first part of the Euromic-Study was published in The Lancet. This part of the study was mainly concerned with trans fatty acids and diseases of the coronary vessels. The second part of the study was related to intake of TFA and cancer. Approximately 2 to 3% of the fat of the milk and derivates have a cis configuration being elaidinic acid (C18:1 trans) responsible for most of the trans fatty acids of milk.

In some cases the amount of trans fatty acids in milk and derivates can be as high as 5%. Therefore the International Federation of Margarine Association (IFMA) and the IMACE (Association of Margarine Industry of the EC Countries) index Association of the Margarine Industry of the EC Countries (IMACE) recommended to lower the amount of TFA (Trans Fatty Acids) in margarine to a maximum of 5% of fat.

Margarine to be used as spread or cooking should have less than 1%. The trans fatty acids can be reduced during refining by reducing the temperature of the process. This however means a longer time the oil or fat has to spend in the system an therefore higher operation costs. The increased costs should be accepted in change of a healthy product.

The deodorization of vegetable oils during refining using temperatures over 230 C results more than 3% of trans fatty acids. Using temperatures under 230 C a maximum of 0,5% trans fatty acids are formed. This can be tolerated. A real great amount of trans fatty acids are formed by partial hydrogenation of soy oil.

A melting point of 36 to 37 C of soya oil very commonly used in the production of margarine as well as deep frying fat and products for bakery results in up to 50% of trans fatty acids.

Trans fatty acids are therefore hidden in cakes, cookies, creams and margarine and all kind of fried product. In USA partially hydrogenated oil is used almost everywhere. The trans fatty acids represents therefore a great menace to health because they act as saturated fatty acids and may cause arteriosclerosis. Trans fatty acids can be avoided during industrial processing of oil using entirely hydrogenated oil. If all double and triple bindings of the molecule of the fatty acids are saturated the trans stereoisometric configuration ceases to exist and there is no negative physiological activity left. To obtain the same consistence of partially hydrogenated oil it is necessary to add more liquid oil and esterified the whole compound.

This implies in higher production cost. All efforts to get healthy products should however be made, including the acceptance of a small increase of price of the final product in order to get margarine, fats, creams and bakery products having less than 5% of their fatty acids in trans configuration.

FATS AND MARGARINE FOR BAKERY

Fats and margarine for bakery must have special stability, structure and melting point. Therefore special hydrogenated oils and fats are needed.

Palm oil, hydrogenated melting point 45/46 C	40%
Soybean oil, hydrogenated melting point 36/38 C	35%
Rapeseed oil, liquid melting point 5 C	25%

This fat blend has a content of trans-fatty acids of 20 to 25%.

Melting Points of Fats Found in Nature: Coconut fat melting point 24-28 C Palm oil melting point 37-39 C Palmkernel oil melting point 26-30 C These melting points are to low for the production of pastry margarine.

To avoid hydrogenated oils and fats for bakery it is possible to fractionate fats leaving it at specific temperature to permit the hard components (stearin) to crystallize. Filtration separates the low melting components (olein). Industry still uses hydration instead of fractionated fats because of the the higher prices of fractionated fats. For the sake of health the higher price should be accepted by the consumer in order to get healthy food.

Bakery products bear a lot of hidden fats and are a great source of trans fatty acids with arteriosclerotic and carcinogenic activity.

PHYSIOLOGY OF TRANS FATTY ACIDS

The fatty acids in cis form have the property to reduce blood cholesterol, meanwhile the trans configuration being consumed about 30 to 40 gram a day make the total blood cholesterol and LDL cholesterol to rise. HDL cholesterol is reduced by trans fatty acids.

In countries with high consume of partially hardened soya oil like USA there is an increase of heart diseases like heart infarct.

Industrial manufactured oils and fats cannot avoid completely trans fatty acids. By using proper technologies the content of trans fatty acids however can be reduced to a tolerable amount of maximum 5%.

A daily intake of 4 to 6 gram of trans fatty acids are told to be harmless. All effort should be made by the industry to reduce the amount of trans fatty acids.

TRANS FAT CLAIMS, FDA PROPOSAL

In its November 1999 proposal, FDA proposed a definition for the nutrient content claim "trans fat free" and proposed limits on the amounts of trans fat wherever saturated fat limits are placed on nutrient content claims, health claims, or disclosure and disqualifying levels.

With regard to the specific definitions, FDA proposed that "trans fat free" and "saturated fat free" should be defined as less than 0.5 g trans fat and less than 0.5 g saturated fat per reference amount and per labelled serving; "low saturated fat" as 1 g or less of saturated fat and less than 0.5 g of trans fat per reference amount and not more than 15 percent of calories from saturated fat and trans fat combined; "reduced saturated fat" as at least 25 percent less saturated fat and at least 25 percent less saturated fat and trans fat combined; "lean" as 4.5 g or less of saturated fat and trans fat combined; and "extra lean" as less than 2 g of saturated fat and trans fat combined.

In addition, cholesterol claims were allowed only on foods containing 2 g or less of saturated fat and trans fat combined, and disqualifying and disclosure levels were set at 4 g or less of saturated fat and trans fat combined. FDA did not propose to define "low trans fat."

Quantitative Declaration of Trans Fat in the Nutrition Facts Panel

The Food and Drug Administration (FDA) amended its regulations on nutrition labelling to require that trans fatty acids be declared in the nutrition label of conventional foods and dietary supplements on a separate line immediately under the line for the declaration of saturated fatty acids.

FDA is revised Sec. 101.9(c) by adding paragraph Sec. 101.9(c)(2)(ii) to require the quantitative declaration of trans fat in the Nutrition Facts panel. This new paragraph requires the listing of trans fat on a separate line under the statement for saturated fat.

As is the case for all subcomponents of total fat, it is to be indented and separated by a hairline, with the amount expressed as grams per serving to the nearest 0.5 g increment below 5 g and to the nearest gram increment above 5 g. If the serving contains less than 0.5 g, the content must be expressed as 0, except when the statement ``Not a significant source of trans fat" is used. In addition, the agency is clarifying that the word "trans" may be italicized to indicate its Latin origin.

Not a significant source of...

Section 101.9(c) requires that information on mandatory nutrients, such as saturated fat and trans fat, be included in all nutrition labelling unless otherwise excepted from such labelling as provided for in specified paragraphs.

Special provisions within Sec. 101.9(c) allow for shortened formats that provide manufacturers flexibility to omit noncore nutrients (i.e., mandatory nutrients other than calories, total fat, sodium, total carbohydrate, and protein) that are present in insignificant amounts from the list of nutrients and group them in a summary statement at the bottom of the label that states-Not a significant source of------(see 58 FR 2079 at 2083, Comment 8, January 6, 1993).

These special provisions are found in Sec. 101.9(c)(1)(ii) for calories from fat, Sec. 101.9(c)(2)(i) for saturated fat, Sec. 101.9(c)(3) for cholesterol, Sec. 101.9(c)(6)(i) for dietary fibre, Sec. 101.9(c)(6)(ii) for sugars, and Sec. 101.9(c)(8)(iii) for vitamin A, vitamin C, calcium, or iron.

For consistency with the labelling scheme for these other noncore mandatory nutrients, new Sec. 101.9(c)(2)(ii) provides that if the trans fat content is not required and, as a result, not declared, the statement-Not a significant source of trans fat-must be placed at the bottom of the table of nutrient values.

PHYSIOLOGY OF SATURATED FATTY ACIDS

Saturated fatty acids rise the blood level of LDL cholesterol.

U.S. Food and Drug Administration has approved to label foods containing at least 6,25 grams of soy protein per serving touting a link between eating soy and lower risk of heart disease. 6,25 grams of soy proteins are one-fourth of the 25 grams of soy protein daily which are supposed to be needed to show a significant cholesterol-lowering effect.

The claim was requested by Protein Technologies International, a subsidiary of DuPont Co, which is a manufacturer of isolated soybean protein.

Foods which may be qualified for this claim are soy beverages, tofu, soy-based meat alternatives and some baked goods.

OMEGA-3 FATTY ACIDS CLAIMS

The major source of omega-3 fatty acids is dietary intake of fish, fish oil, vegetable oils (principally canola and soybean), some nuts such as walnuts, and, dietary supplements.

The Oxfor-Durham Study Concerning Omega-3 Fatty Acids and DCD

Alexandra J. Richardson and colleagues studied the effect of dietary supplementation with omega-3 and omega-6 fatty acids on children with developmental coordination disorder (DCD). This disorder affects 5% of school-aged children. They present deficits in motor function associated with difficulties in learning, behaviour, and psychosocial adjustment. The authors found significant improvements in reading, spelling, and behaviour, however, no effect of treatment on motor skills was apparent. They concluded that fatty acid supplementation may offer a safe efficacious treatment option for educational and behavioural problems among children with DCD.

The British company Dairy Crest relying on these results claimed that her omega enriched milk could enhance children's ability to concentrate and learn. The Britain's Advertising Standards Authority requested to stop this claim alleging that children would have to drink more than five litres of that milk every day to get the same amount of omega-3, being thus misleading.

Britain's Joint Health Claims Initiative (JHCI) has approved a generic health claim that foods containing omega-3 benefit heart health, but not learning ability or concentration. The JHCI offers pre-market advice and a code of practice for the food industry, enforcers and consumers, to ensure that health claims on foods are both scientifically truthful and legally acceptable.

UK Food Standards Agency, concerning omega fatty acids said that there is insufficient quality evidence to reach firm conclusions on the effect of nutrition and dietary changes on learning, education or performance for all schoolchildren.

The Agency maintains their advice to a diet lower in fat, sate and sugar but high in fruits, vegetables and complex carbohydrates, in addition to being physically active.

Omega-3 Fatty Acids and Brain Function

Omega-3 fatty acids were proposed as having an important role in mental health, because up to 60% of the adult brain is composed of lipids (dry weight). Thirty five percent of the lipids are phospholipids comprised of unsaturated fatty acid such as docosahexaenoic acid (an omega-3 fatty acid) and arachidonic acid (an omega-6 fatty acid) acids.

Omega 3 and Psychiatric Disorders

Disorders of mental health are becoming increasingly common in the US. It is estimated that in a given year, 22%, or one in five American adults, suffers from a diagnosable mental health disorder.

These disorders, including major depression, bipolar disorder, schizophrenia, and obsessive-compulsive disorder, account for four of the ten leading causes of disability in the US and other developed countries. Many people suffer from more than one mental disorder at a given time.

Schachter, and colleagues co authors of the Oxford-Durham study write that overall, other than for the topics of schizophrenia and depression, few efficacy or safety studies were identified.

Only with respect to the supplemental treatment of schizophrenia is the evidence even somewhat suggestive of omega-3 fatty acidsâ potential as short-term intervention. Additional research might reveal the short-term or long-term therapeutic value of omega-3 fatty acids.

One study demonstrating a significant clinical effect related to 1 g/d E-EPA given over 12 weeks to 17 patients with depressive symptoms cannot be taken to support the view of the utility of this exposure as a supplemental treatment for depressive symptomatology or disorders.

Nothing can yet be concluded concerning the clinical utility of omega-3 fatty acids as supplemental treatment for any other psychiatric disorder or condition, or as a primary treatment for all psychiatric disorders or conditions examined in the review. Primary treatment studies were rare.

Much more research is needed before the possible utility of (foods or supplements containing) omega-3 fatty acids as primary prevention for psychiatric disorders or conditions can be ascertained. Studies of omega-3 fatty acidsâ™primary protective potential in mental health could be œpiggybacked onto longitudinal studies of their impact on general health and development.

FSA Systematic review of the effect of nutrition, diet and dietary change on learning, education and performance of school children carried out by the University of Teeside

The UK's Food Standards Agency has published the results of a systematic review of the effect of nutrition, diet and dietary change on learning, education and performance of school children aged 4-18 years, covering already published studies.

The Government and those involved in education are committed to improving learning and raising standards in schools, as well as meeting the needs of individual pupils.

There is widespread belief that nutrition and diet may have a part to play in this process; however, there is a degree of uncertainty as to what interventions or supplements work.

Much of the available evidence is confusing and contradictory. In essence those charged with supporting and delivering education are seeking clear guidance for both individuals and groups of children, so as to be able to identify what probably works, what might work and what probably does not work.

The authors conclude that there is insufficient evidence to identify any effect of nutrition, diet and dietary change on learning, education or performance of school aged children from the developed world. Further research is required in settings of relevance to the UK and must be of high quality, representative of all populations, undertaken for longer durations and use universal standardised measures of educational attainment.

However, challenges in terms of interpreting the results of such studies within the context of confounders such as family and community context, poverty, disease and the rate of individual maturation and neurodevelopment will remain.

Whilst the importance of diet in educational attainment remains under investigation, the evidence for promotion of lower fat, salt and sugar diets, high in fruits, vegetables and complex carbohydrates, as well as promotion of physical activity remains unequivocal in terms of health outcomes for all school children.

Omega-3 Fatty Acids from Diet or Supplements and Depression

Omega-3 fatty acid supplementation have been linked with behavioural improvements of children with learning difficulties, behavioural problems or Attention Deficit Hyperactivity Disorder (ADHD). According to the phospholipid hypothesis the decreased omega-3 fatty acid intake could be responsible for the disease.

According to the US Department of Health and Human Services 3% of children and 12% of adolescents may suffer from clinical depression. Professor Haim Belmaker and colleagues treated for one month children aged between 6 and 12 with a combination of eicosapentaenoic acid (EPA) and docosahexaenoic acid (DHA), an omega-3 supplement commonly available at drugstores.

The researchers found significant depression reductions in the omega-3 treated group, concluding that Omega-3 fatty acids may have therapeutic benefits in childhood depression.

Cognitive function seems to benefit from DHA which is involved in the membrane of ion channels in the brain, making it easier for them to change shape and transit electrical signals. EPA may influence brain function directly increasing blood flow in the body, affecting hormones and the immune system.

Belmaker and colleagues concluded that omega-3 fatty acids may have therapeutic benefits in childhood depression.

Omega-3 Polyunsaturated Essential Fatty Acid Status as a Predictor of Future Suicide Risk

Low levels of docosahexaenoic acid, a polyunsaturated fatty acid, and elevated ratios of omega-6/omega-3 fatty acids are associated with major depression and, possibly, suicidal behaviour.

M. Elizabeth Sublette and colleagues found in a study that a low docosahexaenoic acid percentage and low omega-3 proportions of lipid profile predicted risk of suicidal behaviour among depressed patients over the 2-year period. If confirmed, this finding would have implications for the neurobiology of suicide and reduction of suicide risk.

Omega-3 Fatty Acids and Mood Disorders

In an overview of epidemiological and treatment studies concerning deficits in dietary-based omega-3 polyunsaturated fatty acids, resulting mood disorders and their therapy with omega-3 fatty acids supplementation Gordon and colleaugues found that according to the different authors reviewed eicosapentaenoic acid (EPA) or docosahexaenoic acid (DHA), is likely to provide the greatest benefit. More studies studies clarifying the efficacy of omega-3 supplementation for unipolar and bipolar depressive disorders are needed. Mood gene D4DR The gene, called D4DR (fourth dopamine receptor gene) and located on the 11th chromosome, is said to be responsible for 10 percent of people's novelty-seeking or adventurous behaviour. The D4DR gene was discovered in the United States in 1991.

Its functioning is expressed in the limbic part of the brain-the section involved in emotions-and binds with high affinity to clozapine, a drug used to treat schizophrenia. Doctors at Soroka Hospital and the Beersheva Mental Health Center and at Jerusalem's Herzog Memorial-Ezrat Nashim Hospital, who tested 124 healthy Israelis, made the discovery. Doctors from the U.S. National Cancer Institute also pinpointed the gene on 300 people of various ethnic groups at the University of Maryland.

People who score low on the novelty-seeking test tend to be exploratory, fickle, excitable, quick tempered, and extravagant, while those who score high are more stoic, loyal, reflective, frugal, rigid, and even-tempered.

An U.S. team, which included Dr. Jonathan Benjamin of Soroka, backed up the Israeli findings.

Changing Nutritional Habits Regarding Omega Fatty Acids

Nutritional habits have changed from whole grains, beans and other seeds, and seafood high in omega-3 fatty acids to prepared foods containing corn oil, safflower oil, cottonseed oil, peanut oil, soybean oil and red meat, which are high in omega-6.

EM Berrry says omega-6 fatty acids are essential for normal growth, development and health, and so extreme care is necessary before deciding that they are harmful The relation n-6: n-3 changed from 3:1 towards 1:20

n-6 function cannot be considered in isolation but needs to be seen as part of the complex of nutrient interactions with n-3 fatty acids (which compete for the same enzymatic pathways) and antioxidants.

Insulin sensitivity might be the common factor relating disease to fatty acid metabolism both within and between the fatty acid pathways.

High linoleate to arachidonate concentrations have been observed in insulin resistance, diabetic complications and some tumours, but these are multifactorial processes that include many lifestyle determinants and it is therefore wrong to condemn only n-6 fatty acids in their etiology.

Omega-6 and Prostate Cancer

New research, leaded by Professor William Aronson found that changing the ration of omega-3 to omega-6 in the typical Western diet might reduce prostate cancer tumor growth rates and prostate specific antigen PSA levels. PSA is a marker for the risk of prostate cancer.

Omega-3 fatty acids(eicosapentaenoic acid (EPA) and docosahexaenoic acid (DHA)) and the omega-6 acid (arachidonic acid) compete to be converted by cyclooxgenase enzymes (COX-1 and COX-2) into prostaglandins, which can become either pro-inflammatory and increase tumour growth, or anti-inflammatory and reduce growth.

Aronson and colleagues found that levels of the pro-inflammatory prostaglandin (PGE-2) were 83 per cent lower in tumours in the omega-3 group than in mice on the predominantly omega-6 fatty acid diet.

According to this study higher levels omega-3 fatty acids may lead to development of more anti-inflammatory prostaglandins. The authors conclude that eating a healthier ratio of these two types of fatty acids may make a difference in reducing prostate cancer growth.

The Role of Fat and Cardiovascular Diseases

The Women Health Initiative Randomized Controlled Dietary Modification Trial leaded by Barbara Howard found that dietary changes like eating less fat and more vegetables has no effect on the risk of cardiovascular disease.

This study is inconsistent with earlier studies which associate dietary fat intake and a lowered risk of cardiovascular disease, focusing on wholegrain, fish oil and omega-3 fatty acids. The authors, however, find it possible that a diet specifically lower in saturated and trans fat combined with increased intakes of vegetables, fruits, and grains might have led to a decrease in cardiovascular disease risk.

According to Cheryl Anderson and Lawrence Appel The Women's Health Initiative study had not considered current dietary guidelines to cut CVD risk, such as less salt and more potassium, the DASH (Dietary Approach to Stop Hypertension) diet, and weight loss.

Current policies towards fruit, vegetables and wholegrain would probably be unaffected.

The study, however, may indicate that saturated fat levels should be lowered from 10 per cent to seven per cent of total energy intake. The possibility that the effect might have been greater in men or if the diet had been initiated at younger ages cannot be ruled out.

Reduction of fat level in fried fish: Three hydrocolloid coating materials were tested in reducing fat uptake in battered fish products.

Camden and Chorleywood Food Research Association Group coatings such as alginate, pectin, gellan gum, methyl cellulose, and hydroxypropyl methyl cellulose.

Fish fillets were coated with alginate, pectin and gellan gum and then fried. All three coated battered fish fillets with either water or one of the three hydrocolloids showed reduced fat level in the final product.

These findings could help to reduce total lipid intake by the population as fried fish is an important constituent of the average population.

The authors suggest that the reduction of fat was due to the waterbinding ability of the coatings. In fully fried products, the reduction in fat uptake was less marked, although an effect was still seen.

It is likely that the longer frying times compromised the integrity of the coatings, emphasising the need to consider the requirements of both product and process when applying edible films.

Green Paper: Promoting Healthy Diets and Physical Activity

The Obesity Crisis: The Green Paper of the Commission of the European Communities entitled" Promoting healthy diets and physical activity: towards a European strategy for the prevention of overweight, obesity and chronic diseases" opens the discussion on the obesity crisis, focusing on the factor of food for the prevention of chronic diseases, overweight and obesity

Improving the health of Europeans through better diets and greater physical activity is crucial to preventing a range of non-communicable diseases and improving quality of life for millions of people.

According to the Confederation of Food and Drink Industry of the EU a greater understanding of all obesity-related factors is needed, calling for improved public health education on nutrition and healthy lifestyles. This would enable consumers to take responsibility for making healthy choices. It will be also be the basis for understanding and making use of product information provided by the industry.

Strategies should include the determinants that affect food choice, factors that lead to insufficient physical activity in everyday life, and not just food products themselves. The Green Paper considers industry self-regulation the best way of dealing with the problem. According to the CIAA, however, a broader approach is needed to meet all factors involved, such as:

- It should be clear that each consumer is responsible for ensuring that his or her own lifestyle is a healthy one.
- Parents have a similar responsibility for their children.
- Increase of level of physical activity in children, adolescents and adults, in particular in the school environment.

The European Vending Association (EVA) questions the scientific data of the Green Paper regarding the excessive intake of energy-dense snacks and sugar-sweetened soft drinks: EVA says the "notion of excessive intake" is vague and very subjective and it is unclear how it impacts directly on the Body Mass Index EVA calls for other factors such as energy out determining if intake is excessive or not.

NEW DISCUSSION CONNECTING HIGH FRUCTOSE SYRUP WITH OBESITY

High fructose corn syrup (HFCS) was introduced in the 1970s. Food industry, particularly the soft drink industry, uses fructose syrup in excess. Removing fructose from soft beverages could help to reduce obesity, as a possible mechanism is suggested which may explain the link between rising obesity and sweetened beverages.

Three important studies report that high fructose corn syrup is an important factor in the rising obesity epidemic:

According to Hella S. Jürgens and colleagues (2005), from the Department of Pharmacology, German Institute of Human Nutrition, Potsdam-Rehbruecke, exposure to fructose water increased adiposity, whereas increased fat mass after consumption of soft drinks or diet soft drinks did not reach statistical significance.

Total intake of energy was unaltered, because mice proportionally reduced their caloric intake from chow. The researchers found that fructose also produced a hepatic lipid accumulation with a characteristic pericentral pattern.

Jürgens conclude that a high intake of fructose selectively enhances adipogenesis, possibly through a shift of substrate use to lipogenesis.

Swiss researcher Kim-Anne Le and colleagues report in December 2006 that moderate fructose supplementation over 4 weeks increases plasma triacylglycerol and glucose concentrations without causing ectopic lipid deposition or insulin resistance in healthy humans.

In February 27, 2007 Juan Carlos Laguna and colleagues wrote that liquid fructose changes the metabolism of fat in the liver by impacting a specific nuclear receptor called PPAR-alpha, leading to a reduction in the liver's ability to degrade the sugar.

According to the authors, this would partly explain the link between increased consumption of fructose and widening epidemics of obesity and metabolic syndrome. In their article the authors conclude that hypertriglyceridemia and the retention of fat in liver induced by fructose ingestion result from a reduction in the hepatic catabolism of fatty acids driven by a state of leptin resistance.

According to this research, the fructose increased fat synthesis in the liver and also acted on the PPAR-alpha receptor(which controls the oxidation of fatty acids) to reduce the degradation of the fructose, and reduces the activity of the hormone leptin which is engaged in the metabolism of faty acids in liver.

ANTIOXIDANTS IN HUMAN NUTRITION

Formation of Radicals: Free radicals are told to start cancinogenic reactions in vivo. Oxygen is the main source of free radicals such as Singulett-O2, superoxydation and the hydroxilradical. In small amounts the radicals are used in many biochemical reactions. The amount of free radicals being however to high the benefit of the radicals turn out to be dangerous to the physiology of the cell ending in Arteriosclerosis and cancer. The organism protects itself from free radicals with the building of an own free radical defence system helped by external antioxidant vitamins.

Antioxidant vitamins	Protective function against free radicals
Carotenoids (provitamin A)	Singulett-O2 quencher.
Tocopherols (vitamin E)	Act as radical receiver in lipid layers together with vitamin C mainly to protect polyunsaturated fatty acids. Tocopherols also protects beta-carotene and vitamin A from autoxidation in cells.
Vitamin C	Act as radical receiver in the cytoplasm together with vitamin E. It also regenerates tocopherol from tocopheryl radicals which were originated during the antioxidant function. Aspirin can triple the rate of excretion of vitamin C
Selenium, Zinc, copper, manganese	As elements being found as traces in enzyme system of the cells. An undersupply of these elements reduces the efficiency of vitamins, therefore they are here included
Magnesium	Magnesium is a light metal which can burn in presence of oxygen, being used as torch but also in metal alloys in aeronautic. It is present in minerals, seawater and mineral water, in plants as chlorophyll and wheat bran 590 mg/100g, slim cocoa powder 500 mg/100g, sunflower seed kernel 20 mg/1000g, sesame seed 350 mg/100g, cashew nut 270 mg/100g peanuts 163 mg/100g, oat flakes instant 140 mg/100g dried figs 70 mg/100g, Edam cheese 45% 59 mg/100g cooked spinach 50 mg/100g, wholemeal bread 92 mg/100g white bread 19 mg/100g. A balanced diet supplies sufficient magnesium It is an important biological element for animals and humans. It is a physiological antagonist of calcium. Values of serum are 0,65-1,03 mmol/l Magnesium is very important for many enzymatic reactions (it activates all reactions where ATP is present. Important for undisturbed building of bones, It acts relaxing. It is being told that magnesium improves the connections of the synapsis of nerve cells having therefore anti stress function, however there is no scientific evidence for it. According to D.

	Hötze, C. Küpper and A. Zittermann research has been done relating a permanent effect of the streß hormones catecholamine and cortisol. The result of this research was that there is a loss of magnesium. Other findings say there is an improving of the stress situation with supplementation of magnesium working in very laud places. An undersupply of Magnesium may be an additional factor of the origin of arteriosclerosis Undersupply of Magnesium electrolyte syndrome caused by undersupply of magnesium due to unbalanced diet, infusion therapy, chronic diarrhea, special diets, excessive alcoholism. laxatives, inflammation of the pancreas, diseases of the intestinal tract. faulty absorption syndrome, magnesium loss through urine with diuretica therapy. chemotherapy, Diabetes mellitus Syndromes of undersupply of magnesium are: Loss of appetite, nausea, vomiting,lack of drive, weakness. Course tremor, cramps, cramp in the leg during the night. Tetanic contractions.
Selenium	Aside of its function in enzymes selenium is known acting by itself as an antioxidant and probably protecting from cancer. All other inorganic elements have not been found with antioxidant properties.

RDI of Vitamin C

The recommended daily intake of the vitamin C in Europe is 60 mg. In the US, men are recommended to consume 90 mg per day, and women 75 mg per day.

To provide data for the RDA of vitamin C, Mark Levine and colleagues conducted an in-hospital depletion-repletion study. They found that bioavailability was complete for 200 mg of vitamin C as a single dose. No vitamin C was excreted in urine of six of seven volunteers until the 100-mg dose. At single doses of 500 mg and higher, bioavailability declined and the absorbed amount was excreted. Oxalate and urate excretion were elevated at 1000 mg of vitamin C daily compared to lower doses. Based on these data and Institute of Medicine criteria, the current RDA of 60 mg daily should be increased to 200 mg daily, which can be obtained from fruits and vegetables. Safe doses of vitamin C are less than 1000 mg daily, and vitamin C daily doses above 400 mg have no evident value.

Antiinflammatory effects of Vitamin C

Blood levels of C-reactive protein (CRP) and tissue plasminogen activator (t-PA), known as markers linked to inflammation were found inversely associated with Plasma vitamin C, fruit intake, and dietary vitamin C.

Goya Wannamethee and colleagues from the Royal Free and University College Medical School, London concluded that vitamin C has antiinflammatory effects and is associated with lower endothelial dysfunction in men with no history of cardiovascular disease or diabetes.

The authors found that the high blood levels of vitamin C were associated with a 45 per cent reduced risk of inflammation (with respect to CRP levels), and high fruit intake was related to a 25 per cent reduced risk of inflammation.

Plasma (but not dietary) vitamin C also showed inverse associations with both fibrinogen concentrations and blood viscosity. No associations were seen with von Willebrand factor or factor VIII.

Critics on the Wannamethee Study

Previous clinical trials (with diabetics, smokers and healthy men) had not reported an anti-inflammatory effect from vitamin C supplementation.

In contrast, intravenous vitamin C trials did report an improvement in endothelial function.

Ishwarlal Jialal and Uma Singh from the University of California Davis Medical Center, writes in an editorial that in respect to the antiinflammatory effects of vitamin C, the article of Wannamethee does not allow the drawing of any valid conclusions. Much further research in a dose-response structure is required to ascertain whether oral vitamin C supplementation is antiinflammatory and whether it improves endothelial dysfunction. Until such studies have been conducted, it is safe to adhere to the guidelines of national organizations to consume e 5 or more daily servings of fruit and vegetables.

Other shortcomings of the article of Wannamethee are cited in this editorial: The study was only focused on elderly white men and thus could not be generalized for other groups.

The use of t-PA as a measure of endothelial inflammation is being questioned.

Previous clinical trials (with diabetics, smokers and healthy men) had not reported an anti-inflammatory effect from vitamin C supplementation. Intravenous vitamin C trials did report an improvement in endothelial function.

The authors call for more research in a dose-response to ascertain whether oral vitamin C supplementation is anti-inflammatory and whether it improves endothelial dysfunction.

Different Forms of Vitamin E

There are eight forms of vitamin E: Tocopherols: alpha, beta, gamma, delta. Alpha-tocopherol is found in supplements and in the European diet, and gamma-tocopherol is found in the American diet.

Tocotrienols: alpha, beta, gamma, delta. They are found in palmoil, cereal grains and rice bran.

Tocotrienols Stopping the Spread of Cancer Cells

Yoshiyuki Mizushina from the Kobe-Gakuin University, leading author studied the effects of all eight forms of vitamin E on the inhibition of mammalian DNA polymerase, the enzyme that assists DNA replication.

He found that tocotrienols could stop the spread of cancer cells.

The four tocopherols did not influence the activities of mammalian polymerases and had no effect on the spread of cancer cells.

Alpha-and delta-tocotrienols inhibited polymerase lambda activity, and inhibited the spread of cancer cells, and angiogenesis (the growth of new blood vessels) is inhibited and the spread of the cancer to other parts of the body is hindered reducing risk of metastasis.

Other Study by Chandon Sen, Savita Khanna and Sashwati Roy support identical claims for tocotrienols, including neuroprotection, reduction of cholesterol, as an antioxidant, and other anti-cancer studies.

ACTIVITY OF POLYPHENOLS

These substances interrupt the radical chain reaction in lipids of the cell membrane avoiding the formation of oxidized LDL which starts ateriosclerotic diseases. They have reducing properties. Flavonoids may also act as antimutagenic and anticarcinogenic.

Some activities of flavonoids in human are studied but are not confirmed yet:

Inhibition of enzymes with carcinogenic activity.

Inhibition of proteolytic enzymes.

Polyphenoles of these groups are found in many fruits and vegetables.

That is why more fruits and vegetables and less meat, fat and alcoholics should be consumed.

Tea is the main source of catechins. According to the Rotterdam-Study the consumption of tea protects against heavy aorta ateriosclerosis in particular in women.

Important varieties and their content of polyphenoles are:

Camellia sinensis var. assamica:

- High content of catechins and coffein
- High activity of Phenoloxidase activity resulting in dark brown colour of black tea

Camellia sinensis var. sinensis:

- Lower content of catechins and caffeine as noted in assamica

- Lower activity of Phenoloxidase activity which makes it suitabler for the production of green tea

Protective effect of green tea and soy intake in relation to breath cancer risk

There is substantial in vitro and in vivo evidence implicating tea polyphenols as chemopreventive agents against various cancers.

However, epidemic data collected by researchers of the Department of Preventive Medicine, University of Southern California, Keck School of Medicine, Los Angeles are not supportive of a protective role of tea, mainly black tea, in the aetiology of breast cancer.

The case study showed that both green tea and soy intake had significant, independent protective effects on breast cancer risk The results of this case study point to an important role of both green tea and soy intake in relation to breast cancer risk.

Anticancer Effects of Green Tea

Shaun K. Rodriguez et col. suggest a novel mechanism for green tea catechin, epigallocatechin-3-gallate anticancer effects where epigallocatechin gallate can abrogate vascular endothelial growth factor signalling by interfering with the formation of a receptor complex, resulting in attenuated mitogenic and angiogenic signalling.

Bitter chocolate has high levels of catechins (53,3 mg/100g chocolate). Milk chocolate has only 15,9 mg catechins/100g chocolate. Catechines are situated in the cocoa part of the cocoa beans (Theobroma cacao), that is why bitter chocolate with high cocoa is rich in catechines.

Analytical test for total polyphenols is the Folin-Ciocalteau. Due to the reducing activity of the polyphenols a strong blue colour is created which can be be measured.

Specific polyphenols are measured with aid of HPLC or capyllary electrophoresis.

In order to act reducing it is necessary have at least one of the following structures:

1. 3',4'-Dihydrofunction at the B ring. This makes the turnover of a proton building thus an aryloxy-radical.
2. OH-A group in position 5 and 7 at the A ring.
3. A doublebound in 2-3 position in conjugation with the 4-Oxo-function and the 3-OH group at the ring C turning possible the delocalization of the electrons of the B ring.

Quercentine has all three structures turning out to be an excelent antioxydant in water solution.

Flavonols in Chocolate

A 15 years follow up study made by Brian Buijsse and colleges indicate that cocoa-containing foods improve endothelial function and reduce blood pressure. In a cohort of elderly men Brian Buijsse and colleges found that cocoa intake is inversely associated with blood pressure and 15-year cardiovascular and all-cause mortality, confirming previous studies which have linked flavanols (Flavan-3-ols) to improved cardiovascular health.

The cocoa-containing foods from this study included chocolate confectionery, cocoa sandwich filling, cocoa desserts, cocoa drinks, and dietary supplements.

The men who consumed the most cocoa (more than 2.3 grams per day) had lower systolic diastolic blood pressures (3.7 mmHg and 2.1 mmHg, respectively) than those who consumed the least cocoa (less than 0.36 grams per day).

Several previous studies have shown that flavonol consumption increased blood vessel vasodilation, and improve endothelial function.

However, Cathy Ross, of the British Heart Foundation, said that here is some evidence that when eaten in small quantities, dark chocolate might have some beneficial effects on blood vessels and lowering blood pressure, but as yet no study has investigated the long-terms clinical effects.

People would have to eat about 100 grams of dark chocolate a day with 500 calories and 30 % of fat to get an effective amount of flavanols. According to Cathy Ross there are much better ways of improving heart health like products with increased flavonol content.

LEGUMIN AND HOMOLOGOUS PROTEINS

Legumin in Maize

Legumin is a member of a family of storage proteins (11S globulins) found in the Leguminaceae and other higher plants.

It is a histidine-and glutamine-rich polypeptide of 50â•"55 kDa with a peptide bond joining the C-terminus of the alfa-subunit (always asparagine) and the N-terminus of the beta-subunit (almost always glycine).

Yamagata and colleagues found that a legumin does exist in maize, that it is uncleaved, that it appears to be localized to small protein bodies essentially identical to those found in legumes, and that it is more abundant in wild type (W64A), than in sweet corn or opaque-2 maize.

The authors discuss the phylogenetic relations between maize legumin and 7S globulin (vicilins).

Convicilin in Pisum and Faba Beans

Boulter and colleagues found in Pisum and faba beans two major families of storage proteins, legumin and vicilin. Legumins are hexameric proteins comprising of 6M gama 60 000 subunits.

Vicilins are a less well-defined group of proteins consisting largely ofM gama 50 000 subunits. Boulter found a third storage protein, convicilin, a vicilin-type protein with vicilin immunological determinants.

According to Boulter proteins homologous to legumin have been found in Arachis, Glycine,Vigna unguiculata, Lupinus, Cicer, Lens and Lathyrus. Equivalent vicilin-type proteins occur in Phaseolus vulgaris, Glycine, Arachis and Vigna unguiculata.

Legumin Type Protein in Almond

According to Shridhar and colleagues the almond major storage protein, amandin,Amandin is a legumin type protein. Amandin is composed of two major types of polypeptides linked via disulfide bonds. Amandin is a storage protein Amandin is not a glycoprotein. Amandin-1, amandin-2, and amandin-3 are antigenically related and have similar biochemical properties. Shiddhar found that methionine is the first essential limiting amino acid in amandin followed by lysine and threonine.

Western Blot Legumin Protein

The classical three-step procedure of Western Blot (blocking, primary antibody-binding and secondary antibody-binding) takes about 4.5 hrs.

GenScript developed the One-Step Western which performs Western blot or Dot blot to detect any protein including Legumin protein and related proteins within an hour.

After the proteins are transferred from a gel to the membrane, the membrane is incubated in Pretreatment Solution for 5 minutes and in WB solution with primary antibody added for 40 min, the membrane is washed three times for 5 min each and the blots are develop using either Chemiluminescent substrate or TMB.

DIETARY FATS, BODY FATS AND BLOOD LIPIDS

To a large extent, you are what you eat when it comes to dietary fats. Our bodies can manufacture fats from other substances in the diet when there are excess calories. However, on balance, fat provides many of the excess calories in the American diet. In this section, you will learn not only how the excess fat comes into the diet, but how different fatty acid sources have different effects on physiological processes. Fats and oils provide the most concentrated source of calories of any foodstuff. Fats provide essential fatty acids (linoleic and linolenic acids) which are precursors for prostaglandins. Fats also require prolonged digestion and contribute to satiety, carry fat-soluble vitamins, and concentrate the tastes of foods to make them more palatable.

Ninety-five percent or more of the fats you eat and store are triglycerides. These fats have a three carbon backbone and three fatty acids esterified at each of the three positions. When all the fatty acids on a triglyceride are the same they are called simple, otherwise they are called mixed triglycerides, which are more common. Excess calories, regardless of source, are stored as triglycerides.

The principal dietary sources of fat are meats, dairy products, poultry, fish, nuts, and vegetable oils and fats used in processed foods. Vegetables and fruits contain only small amounts of fat, so that vegetable oils are only sources of fat due to processing of vegetables.

The most commonly used oils and fats for salad oil, cooking oils, shortenings and margarines in the U.S. include soybean, corn, cottonseed, palm, peanut, olive, canola (low erucic acid rapeseed oil), safflower, sunflower, coconut, palm kernel, tallow and lard. These oils contain varying compositions of fatty acids which have particular physiological properties.

Comparison of Dietary Fats

Dietary Fat	Saturated Fat	Linoleic Acid	Alpha-Linolenic Acid	Monousaturated Fat
Corn Oil	13%	61%	1%	25%
Olive Oil	14%	8%	1%	77%
Soybean Oil	15%	54%	7%	24%
Peanut Oil	18%	34%	0%	48%
Safflower Oil	9%	78%	trace	13%
Sunflower Oil	11%	69%	0%	20%
Palm Oil	51%	10%	0%	39%
Lard	41%	11%	1%	47%
Beef Tallow	52%	3%	1%	44%
Butterfat	66%	2%	2%	30%
Coconut Oil	92%	2%	0%	6%

Reference: Agricultural Handbook No.8-4 and Human Nutrition Information Service, USDA, 1979.

In starvation and overfeeding, the body regulates the metabolism of carbohydrates and protein closely, but allows the stores of fat to expand easily in overfeeding and to contract with underfeeding. Since fat yields 9 kcal/gm and requires little water for storage, it is a very efficient store of calories. In the non-obese 70 kg. mythical man, 13.5 kg. of fat will carry 130,000 to 160,000 kcal. while 13.5 kg. of muscle will carry only 54,000 kcal. Utilizing the fat stores, and sparing protein stores is essential to surviving starvation. Since many populations have been exposed to epidemics of starvation and abuse, the tendency to retain fat stores is relatively common and is inherited polygenically with a strong environmental influence. The lipoprotein carrying cholesterol, called apoprotein B is one of the largest proteins in the body with over 1300 possible phenotypes, many of which can affect cholesterol metabolism and transport. The common occurrence of hypercholesterolemia in our population is both a result of environmental influence and common polygenically inherited variants of normal cholesterol homeostasis.

1. Lipids-Definition

Fatty acids	Here structure will be important to function with different properties for saturated, unsaturated, monounsaturated, n-3, n-6, etc.
Glycerol containing lipids	Mono, di, tri acylglycerols phosphoglycerides-such as phosphatidyl serine, and lecithin = phosphatidylcholine
Sterols	Also called steroid alcohols-Cholesterol is the primary animal sterol, while vegetable sterols are termed phytosterols e.g. sitosterol and stigmasterol.
Sphingolipids	These special classes of lipids can have special functions within the cell membrane such as sphingomyelin. They are found in high concentration in the brain.
Eicosanoids, Leukotrienes, Prostaglandins	These are involved in cellular signaling and immune function.

Lipids are defined based on their solubility in organic solvents rather than on their structure unlike proteins and nucleic acids. The lipids can be divided into amphipathic and hydrophobic subclasses. Fatty acids and phospholipids are examples of amphipathic lipids in that they dissolve in both aqueous and organic solvents. Examples of hydrophobic lipids are cholesterol,and triacyglycerols.

2. Minor Constituents of Dietary Fats and Oils

Tocopherols	Minor constituents of vegetable oil, they serve as antioxidants. Among the tocopherols, alpha d-tocopherol has the highest vitamin E activity and the lowest antioxidant activity. The antioxidant acitivities of the other tocopherols in descending order is delta, beta or gamma, and alpha. These usually are removed in the processing of vegetable oils but can be added back to improve stability. Tocopherols are not present in appreciable amounts in animal fat.
Carotenoids and Chlorophyll	These are naturally occurring colored substances in fats and oils. Carotenoids range in color from yellow to deep red and have antioxidant activity. Chlorophyll can sometimes tinge oils green, but in most cases it is removed in processing.
Fat soluble vitamins	Fats are poor sources of these except for vitamin E. Vitamins A and D are sometimes added to dairy products, and beta carotene is added to margarine to provide a yellow color.

3. Brief Overview of Metabolism of Fats and Oils

Since triglycerides are the primary ingredient of fats and oils, the digestion and absorption of triglycerides is central to understanding fat absorption and metabolism. Most lipids are not broken down in the mouth, but are mixed with carbohydrates and protein to form chyme in the stomach. This mix is slowly emptied into the duodenum where lipids stimulate cholecystokinin (CCK) secretion. The CCK slows down gastric emptying time to allow more time for intestinal lipid digestion. CCK causes secretion of pancreatic lipase, which breaks down triglycerides in the alkaline environment of the intestine. CCK stimulates pancreatic exocrine secretion of bicarbonate. CCK also stimulates the gallbladder to contract and release bile salts (e.g. glycocholate-derivative of cholesterol) and lecithin (phosphatidylcholine). This is quickly hydrolyzed so that the fatty acids are removed, forming lysolecithin which helps emulsify fats along with bile salts for digestion. Pancreatic lipase works at the oil/water interface since triglycerides are insoluble. Emulsification increases the surface area of the oil /water interface, promoting the breakdown of triglycerides by pancreatic lipase.

The pancreatic lipase preferentially hydrolyzes the 1 and 3 positions forming beta monoglyceride and two fatty acid moieties. As the process of digestion proceeds, micelles form containing beta-monoglyceride, fatty acids, and bile salts. These amphiphilic spheres have polar surfaces and hydrophobic interiors. They can incorporate fat soluble vitamins, but do not contain triglycerides.

The micelles are in equilibrium with the constituent substances and move to the vicinity of the intestinal cell where the monoglyceride and fatty acid (but not the micelle) are absorbed. The bile salts are not absorbed but take part in an enterohepatic circulation.

In the intestinal cells, the monoglycerides and fatty acids are recombined to form triglycerides and incorporated into chylomicrons for transport in the blood stream.

If the fatty acid chain length is less than 10 (such as medium chain triglycerides), then they are transported by the lacteals directly to the liver. In the hepatocyte these short chain length fatty acids can enter the mitochondria without any special transport and are oxidized for energy. On the other hand, the majority of the triglycerides in the diet have a longer chain length and require transport in the bloodstream by chylomicrons and require carnitineacyltransferase action to move into the mitochondrion for oxidation.

The body can make saturated and monounsaturated fatty acids by modifying other fatty acids or by de novo synthesis from carbohydrate and protein. However, the polyunsaturated fatty acids linoleic and linolenic are essential fatty acids and must be supplied in the diet. Needless to say, the problem of fatty acid deficiency is not widespread in the free-living population in this country. The minimum intake of linoleic acid is said to be 3% of total calories. Since linoleic acid makes up about 60% of corn oil fatty acids, most diets should contain over 10% of total calories as fat. The average American eats between 35 and 40 percent of total calories as fat. High fat diets eaten ad lib account for as much as 600 excess kcal/day compared to low fat ad lib diets. Dietary guidelines for reduction of heart disease and cancer mortality recommend 30% of total calories as fat.

Fat moblized from fat stores (as during a calorie deficit) is released into the blood as free fatty acids. Lipoproteins carrying exogenous or endogenously synthesized triglycerides are broken down by lipoprotein lipase at the endothelial surface of blood vessels and the fatty acids released are taken up into the adipocyte for resynthesis and storage as triglycerides.

The reverse process of release of stored fat is under the regulation of hormone-sensitive lipase which releases free fatty acids and glycerol. The free fatty acids released associate with albumin and other proteins and are carried throughout the body. The free fatty acids can also be taken up by the adipocytes again for reutilization. When lipolysis is measured experimentally, glycerol release is followed since it is not taken up by the adipocyte after release.

DIETARY FATS: PRACTICAL CONCERNS

Fat in the diet is one of the major contributors to taste and texture. The aroma of sizzling bacon or a steak on the grill comes primarily from the heated fat. The texture of foods is affected by fats, too-butterfat makes ice cream creamy, and fats in baked goods act as shortening (by shortening the strands of gluten in the flour) making the product more tender.

But, since fat is such a concentrated energy source, and because there are other health implications to a high fat diet, food technologists are scrambling to find fat replacements that provide the texture and "mouth feel" of fat without the calories. Some fat replacements are protein-based and some are carbohydrate-based, and therefore yield fewer calories per gram (4 vs. 9), and some are fat-based but are not absorbed by the body.

Understanding the food label is an important skill, since some foods can appear low in fat, and yet still provide 100% of their calories from fat. Lowfat margarine, for example, has only 5 calories per tablespoon, but all 5 calories are fat calories-so the product is 100% fat.

Fat should be expressed as a percentage of calories, not as a percentage of the weight of the food. Lean ground beef labeled "7% fat" sounds lean, but the 7% is an expression of the amount of fat by weight. Since much of the weight of the meat is water (which has no calories), the true percentage of fat is grossly understated. Reading labels carefully allows you to find hidden fats in foods, and to understand why some advertising claims are misleading.

8

THE PHARMACOLOGY OF NUTRITION

AYURVEDA: AN INTRODUCTION

Ayurveda is a Sanskrit word derived from two roots: *ayur*, which means life; and *veda*, which means knowledge. It has its roots in ancient vedic literature. Ayurveda, a system of diet, healing and health maintenance, is probably the oldest science of life, just like the science of Yoga.

Ayurveda includes all aspects of daily life like:

- Maintenance of health
- Prevention of disease
- Harmonization of body and mind
- Natural treatments
- Holistic and complimentary healing techniques

Ayurveda combines yoga, meditation, food, natural preparations, cleansing and regenerative treatments. The overall effect is physical strength, better health, mental clarity, inner peace and calmness.

Ayurveda is a science of life so to know more about it, we must know what is life. Life, according to Ayurveda, is a combination of senses, mind, body and soul. It is clear from this definition of life that Ayurveda is not only limited to body or physical symptoms but also gives a comprehensive knowledge about spiritual, mental and social health.

ROLE OF AYURVEDA

Ayurveda teaches us to understand our body; our particular nature; and our individual mixture of elements at a deep physical, mental and emotional level. With that knowledge we are able to identify activities, conditions, herbs and foods that either keep us healthy and in balance, or make us ill and throw us out of balance.

AYURVEDA AND HEALTH

Ayurveda forms an integral part of the daily regimen of hundreds of millions of people worldwide. Its principles are utilized not only to treat persons who are ill but also to prepare a balanced meal and to construct a harmonious environment to live in.

Ayurveda brings to life the concepts of preventive health care and health promotion.

The goal of Ayurveda is to help an individual discover the knowledge of living and health. Health is the state of harmonious chemical balance in a living organism. Our health depends on the chemical environments inside and outside our bodies. Food plays an important role in creating the internal chemical environment.

THE FIVE ELEMENTS

Our very existence is the interplay of different frequencies of vibrations, which have been classified into a system of five elements by the Ayurvedic seers and sages. These five elements are the agents of the primary inertia principle of consciousness and are the materialized form of the universal energy.

Akasha, or Ether, evolved first and is the subtlest of the five elements. From it comes air, from air evolves fire, from fire evolves water and from water comes earth. The human body, which is composed of these elements, is also nourished and maintained by them.

THE DOSHAS

The starting point for many people into the ancient scientific art of Ayurveda is the relationship of the three Doshas: Vata, Pitta and Kapha. Ayurveda sees life as a harmonic flow, a dynamic balance of those three fundamental forces:

- Vata ? (wind, air) the principle of movement and impulse
- Pitta ? (bile, fire) the principle of assimilation and transformation
- Kapha ? (mucus, water) the principle of stability

These forces act in everyone. When they are in balance they bring well-being and health, in imbalance they lead to feeling unwell and later disease. Everybody is unique and Ayurveda respects this uniqueness. That is why there are individual constitution types, Doshas, in the body. Out of the three basic forces can seven categories be formed:

1. Wind dominated individuals (vata)
2. Bile dominated individuals (pitta)
3. Mucus dominated individuals (kapha)
4. Wind and Bile dominated individuals (vata and pitta)
5. Wind and Mucus dominated individuals (vata and kapha)
6. Bile and mucus dominated individuals (pitta and kapha)
7. Wind, bile and mucus dominated individuals (vata and pitta and kapha in equal proportion)

THE THREE CONSTITUTIONAL TYPES: THE GUNAS

There are three basic constitutional types. To live a long and healthy life, Ayurveda states it is necessary for us to recognize them in ourselves and be able to preserve their balance with nature. It is possible, though, to take a step back and study the mind as Ayurveda clearly tells us that we create what our minds dictate. So do we have any control over our thoughts? Ayurveda says yes, but not in the sense of thought manipulation. We are told that the mind has three tendencies or qualities.

- Satva, the principle of contentment, joy, peace and harmony.
- Rajas, the principle of energy, of change, emotion and turbulence

- Tamas, the principle towards rest, dullness, inertia, depression and resistance

To progress thru life in a harmonious way, it is suggested that we increase Satva and keep a pinch of Rajas and Tamas to accommodate movement and rest. All three Gunas enhance the qualities of the three Doshas.

- A mind tendency toward Satva will encourage the qualities of intelligence, perception and strong digestion found in the Pitta Dosha.
- If the mind of the Pitta type is Rajasic, restlessness, irritation, heat and anger will be dominant, and,
- eventually, through exhaustion, tipping over into dullness and depression. Which are the signs of Tamas.

To help ourselves live a joyful and fulfilled life, we can become aware of our lifestyle and use tools such as breathing techniques, asanas and meditation to calm and integrate the nervous system. Being with like-minded people helps, and of course will food play a significant role.

AYURVEDIC FOOD

The food is either satvic, rajasic or tamasic according to its character and effect upon the body and the mind. One can find out the nature or temperament of a man from the nature of the food he prefers.

Satvic Food

- Satvic foods are fresh, juicy, light, unctuous, nourishing, sweet and tasty.
- It increases the energy of the mind and produces cheerfulness, serenity and mental clarity.
- satvic food is highly conducive to good health.
- A satvic man relishes juicy food and other foods which are attractive in form, soft to touch and pleasant to taste, which are small in bulk but great in nourishment like the words from the lips of a spiritual preceptor.

- Milk, butter, ghee (clarified butter), fresh ripe fruits, almonds, dates, green gram, sprouts, barley, wheat, cereals, tomatoes, plantains etc., are satvic.

Rajasic Food

- Rajasic foods are bitter, sour, salty, pungent, hot and dry.
- These foods create sensuality, sexuality, greed, jealousy, anger, delusion, fantasies, egotism and irreligious feelings.
- The rajasic man always plans to prepare various kinds of preparations to satisfy his palate. The palate remains unsatisfied until the stomach is completely filled with pungent things and till the tongue is burnt with chillies.
- Pungent condiments, meat, fish, eggs, sweets, fried bread, curd, egg plant, carrots, black gram, onions, garlic, lemon, red gram, tea, coffee, betel leaves, tobacco are rajasic articles of food.

Tamasic Food

- Tamasic food is stale, dry, bad smelling, distasteful or unpalatable.
- It increases pessimism, ignorance, lack of common sense, greed, laziness, criminal tendencies and doubt.
- The tamasic man will eat, in the afternoon, food which has been cooked on the previous day and also will like the food which is half-cooked or burnt.
- Foods that have been processed, canned or frozen are tamasic.
- Beef, fish, eggs, wine, garlic, onions and tobacco are tamasic foodstuffs.

Ayurveda places responsibility firmly on the individual to be aware of his or her relationship with nature and she gives us the tools to correct imbalances before they develop into full-blown disease. Sri Kalki says that it is essential to have a healthy body free of mucus in order to attain the state of altered consciousness of enlightenment. A deeper knowledge about Ayurveda will help sincere seekers to understand their bodies and minds from the

Annamaya Kosha's point of view so that it helps in the enlightenment process. The science of Ayurveda, like the science of Yoga, was inspired and developed by the great masters and seers of ancient India. The origin of Ayurveda and Yoga are common to play a highly complimentary role in spiritual evolution and the maintenance of physical well-being and vitality.

Ayurveda is, perhaps, the oldest science of life, a system of diet, healing and health maintenance that is deeply spiritual in origin. Unlike traditional Western medicine, Ayurveda is not confined to healing of disease in a superficial treatment of symptoms. Instead, it evaluates the complete body mind of the individual. Ayurveda sees medicine and diet as complementary rather than separate. No one can expect to retain vitality, recover from disease, or succeed in the practice of Yoga without the appropriate knowledge of the powerful effect diet has on physical health, mental clarity and spiritual progress.

Indeed, yogis place great emphasis on diet as an integral part of the successful practice of any spiritual discipline. Ayurveda addresses not only healing, but also prevention and maintenance of vitality, as crucial in the practice of Yoga. Ancient seers describe the human body and the body of the universe as composed of prana, the primal energy, vital for life and which manifests in the form of earth, water, fire, air and ether.

Any imbalance of these elements in our body is experienced in illness, discomfort or pain. These elements are kept in harmony by a healthy body that consumes them through breath, food, vegetables, seeds, beans, herbs and roots as vital carriers and balancers for the energy of prana in the body. The power of these foods manifests only when they are used in proper combinations and in complete coordination with unique conditions of each individual. This is where the profound effects of Ayurvedic food reveal themselves.

DISCOVERING YOUR CONSTITUTION

The particular combination of energies present when we enter this world at birth is known in Ayurveda as our innate constitution or prakruti. What we do with our constitution is up to us. It is the source for our health, vitality and well-being. It can also be the springboard toward difficulties, if we ignore its needs. The five elements come together to create the three basic constitutional types. They are Vata, Pitta and Kapha. If air and ether predominate in your constitution, you are Vata in nature. If fire and water take the lead in your physical make up, Pitta is the result. If water and earth hold ascendancy in your body at birth Kapha is your constitution.

What does understanding your constitution mean, practically? Each constitution has different needs. Meeting these needs assures balance and better chances for good health and peace of mind. It also offers you more informed choices in what health care methods are most appropriate for you.

How can you discover your constitution? Ideally, a skilled Ayurvedic physician can assess your pulse and give you information about your prakruti (constitution). From pulse examination he or she can ascertain what elements are in balance, what is out, and what needs to be done.

The pulse is utilised in Ayurveda in a way similar to Tibetan and Chinese medicine. Working with a talent Ayurvedic practitioner is an invaluable resource. But what if you live in Boise, Idaho, where there is no vaidya? Each constitution has certain characteristics physical, emotional and mental attributes, which let us know what elements are present.

THE THREE DOSHAS

When the three doshas exist harmoniously the result is a sense of balance and health; an imbalance, which might be caused by an excess (vriddhi) or deficiency (kshaya) of any or all of them, results in a sign or symptom of disease. Our physical constitution and our health both depend on the balance and interplay of the tridoshas.

The three doshas are known as Vata, Pitta and Kapha. Each dosha has its own part to play in the maintenance of the body. The rhythmic pattern of these humors fluctuates periodically,

Trans Fatty Acids

Unsaturated fatty acids can exist in two different geometric forms. These are called the cis and trans forms. Unsaturated fatty acids exist naturally in the cis form. During food manufacturing processes these cis fatty acids may be changed to trans fatty acids. Hydrogenation of margarine causes this to occur. It has been suggested that trans fatty acids can increase the risk of heart disease.

Free Radicals

Free radicals are highly reactive molecules which have been linked to both heart disease and cancer. A number of factors, including alcohol, stress and environmental pollutants can increase the generation of free radicals in the body. Polyunsaturated fats can also generate free radicals, especially when exposed to heat or sunlight. Because of this it is suggested that vegetable oils should be stored out of direct sunlight. Mono-unsaturated olive oil is less vulnerable to free radical generation and so is a better choice for frying. Anti-oxidants such as vitamins A, C and E offer protection against free radicals. Fresh fruit and vegetables are rich in these anti-oxidants.

Dietary Sources

Saturated fats are nearly always from animal foods. Meat, eggs and dairy products all contain saturated fats. Lard and suet are saturated fats. Coconut oil and palm oil are vegetable sources of saturated fats. Olive oil is a monounsaturated fat. Polyunsaturated fats are usually from plant sources. The ratio of polyunsaturated to saturated fats in the diet is often called the P:S ratio. Cholesterol is present in all animal foods but not plant foods. Egg yolks and high-fat dairy products are high in cholesterol.

Required Intakes

Currently it is believed that around 42% of energy in the typical British diet is from fat. Dietary advice is to reduce this. The COMA (Committee on Medical Aspects of Food Policy) report advocated that no more than 35% of daily energy requirement should come from fat whilst the NACNE (National Advisory Committee on Nutritional Education) paper recommends a reduction to no more than 30%. Special emphasis is placed on reducing the amount of saturated fat in the diet. Vegetarian diets tend to be lower in fat than omnivore diets.

However, vegetarians consuming dairy products and processed foods high in fat may still be consuming too much. Advice to vegetarians is to keep fat intake to a minimum, avoid high fat dairy products and processed foods containing dairy fats and hydrogenated vegetable fats, and to use olive oil for cooking purposes.

CEREALS

All too often, cereal products are thought of as nothing more than starchy fillers, and indeed, when you consider how some commercial products like cakes and biscuits and breakfast cereals are made from over-refined grains with nearly all the nutrients except the starch extracted, and then loaded with fat, sugar, artificial flavourings etc., you might be forgiven for thinking that's true. But go back to the original, unrefined grains and you have a wealth of nutrients in a small package. Grains have been the staple foods of many civilisations for thousands of years. Wheat, barley, oats and rye in Europe, maize in America, quinoa in South America, rice in the East, and millet in Africa.

Nutrition: Cereals are seeds of plants, usually members of the grass family but there are a few exceptions. They are annuals, that is they have to be planted every year and at the end of the summer, when they have produced ripe seeds, they die down. Like all seeds, cereals are very nutritious because they contain all the nutrients the embryo plant needs to start growing.

Unrefined cereals are valuable sources of proteins, carbohydrates, B vitamins and also contain some fat, iron, vitamin E and trace minerals and are a very good source of fibre in the diet. Some cereals (wheat, barley, rye and oats) contain the protein gluten, which is essential for leavened bread-making. Without sufficient gluten, bread will not rise. People suffering from gluten

- It builds the body and forms its structural skeleton, muscles, organs, ligaments, tendons and skin.
- Kapha lubricates the body, ensuring the smooth function of joints. With strength, stability, and solidity, Kapha is the body in repose.

If kapha predominates in your constitution, you will have a steady, reliable mind. You will take time to learn, and will remember what you have learned. There can sometimes be elements of dullness with a kapha mind: it is usually content not to seek fresh mental stimulation.

Kapha Governs: Lubrication fluids and supply system, mucous membranes, synovial fluid in joints, binding, firmness, heaviness in the body, softness in the body, distribution of heat, strength and stamina, longevity of cells and thus the person, sleep, long-term memory, security.

Factors Aggravating Kapha

- Eating foods that are very cold, heavy, oily like fried food, sweets like sugars, fats, milk products, ice creams and meat.
- Activities like sleeping during the daytime, not doing much physical activity, living in cold and humid climate etc.

Your Constitution According to Kapha

- Medium and well-built frame.
- Gain weight easily, have a hard time to lose.
- Large, attractive eyes with thick eyelashes.
- Thick, wavy hair, a little oily, dark or light.
- Have ideas and opinions but not confident of the same.
- Work well with routine.
- Steady, reliable, slow to change

AYURVEDA: HEALING WATERS

Ayurveda is perhaps the oldest science of life and a system of diet, healing and health maintenance that is deeply spiritual in origin. In this article the healing properties of water will be presented.

Well water, pond water, rain water, there are many types of water described in ayurvedic texts, and each has a therapeutic value, just as food does. Water represents soma, the nourishing cooling quality that is associated with lunar energy. It helps with digestion, cools and balances Pitta dosha, supports Kapha and counteracts the dryness of Vata. It nurtures, lubricates and also detoxifies when it flows out of the body as urine. Water, when properly absorbed by the body, has several healing qualities.

1. Helps to remove fatigue (shramnashana)
2. Enhances glow of skin
3. Prevents constipation
4. Increase stamina
5. Provides satisfaction
6. Helps the heart
7. Helps digestion
8. Cooling
9. Always helpful to the body
10. Easy to assimilate
11. Life-giving
12. Antioxidant

The healing effects of water can be enhanced using ayurvedic methods.

Water for Cleansing: Sometimes people have dry skin and unquenchable thirst even though they drink lots of water. The deeper physiology is not getting enough moist. This occurs when the person's agni is low and mucus blocks the micro channels, which carry water to the cells. In order to cleanse the channels and enhance moisture absorption, ayurvedic texts recommend boiling the water for various lengths of time, creating therapeutic water called ushnodaka. Another method is to add spices or herbs to the water after boiling.

Why it Works: When the water boils, it gets charged with heat and becoming sharper in quality. This sharpness allows it to cleanse the channels and penetrate deeper levels of the physiology. Spice creates an added therapeutic effect by interacting with the water on the molecular level. Spices create different effects on the body through aroma and taste.

It becomes easier for the body to flush out toxins and impurities because of the sharpness of the agni (heat) in the water and because of the sharpness of the spices. Over time, it cleanses the channels so the water is unobstructed as it travels into the body to hydrate the tissues, and travels out carrying waste.

Ancient texts talk about the difference in the rate of absorption of regular water vs. boiled water:

1. Regular water ? takes about six hours if every channel is clear.
2. Boiled and cooled water ? takes about three hours to be absorbed, and helps open the channels
3. Hot herbalized water ? takes about 1 1/2 hours, due to sharpness of Agni, herbs and spices.

Water for Your Body Type: An ayurvedic expert can design a therapeutic water recipe to give a specific benefit. One water recipe might enhance immunity, another might cleanse the skin, and another might help with prostate imbalance. You can also choose a spice-water recipe for your body type of imbalances.

Vata Balancing Water: Boil two quarts of water for 5 minutes. Take it off the heat and add 3 leaves mint, 1/2 tsp ajwain (ajwain is also known as Bishop's Weed, relatid with cumin but tastes more like thyme with an astringent edge) and 1/4 tsp singhada flour. Place the water in a thermos. Sip it throughout the day at a warm but not hot temperature.

Pitta Balancing Water: Boil two quarts of water for 2 minutes. Take it off the heat and add 1/4 tsp ajwain, 2 rose buds and 1 clove. Store in hot in inside a thermos, but before drinking pour it into a cup and let it cool to room temperature in summer. In winter, it can be slightly warmer.

Kapha Balancing Water: Boil two quarts of water for 5 minutes. Take it off the heat and add 3 holy basil leaves (Tulsi), two thin slices of fresh ginger, 1/4 tsp of cumin and 1/2 tsp of ajwain. Place the water and spices in a thermos, and sip the water at a hot or warm temperature throughout the day.

How Much is Enough: How much water you should drink depends on your age, how much physical work or exercise you do, the weather, your diet, your stress levels, your herbal food supplements, and your body type. The warm Pitta types usually are thirstier than the watery Kapha types. Vata types are often constipated or have dry skin and thus need to drink more water.

We recommend making your spice water first thing in the morning and sipping it every fifteen minutes throughout the day. Drink plain water after 7:00 pm, as spice-water is too enlivening to drink right before sleeping. If you don't finish the spice-water by then, throw it out and start fresh in the morning.

You may want to drink some plain water during the day as well. If you have been exercising and need to drink a full glass of water, it's better to drink plain water rather than the spice-water.

AYURVEDIC VIEW ON EATING FOR ENERGY

When you need a high energy fix, do you reach for a power bar or power shake? Or do you resort to caffeine, chocolate, or high-sugar snacks when you feel that end of the day sag? It may surprise you, but according to Ayurveda, the real energy-boosters are fresh fruits, vegetables, species, and whole grains. These are the foods that are rich in chetna, the healing and nurturing influence of nature.

They are so alive with nature's intelligence that fatigue-causing toxins cannot accumulate in the body when you eat them.

Grains for Power

Athletes have long relied on carbohydrates in grains for long-term endurance and energy. Yet not all carbohydrates are created alike. A croissant, for instance, is high in fat and low in nutrition.

The most nutritious carbohydrates are whole grains, which have been found to lower cholesterol inhibits cancer causing activity and harmful bacteria in the intestines, and lower blood glucose level. And these whole grains, ayuerveda considers rye, quinoa, amaranth and millet as the most nutritious, because they are especially high in protein and mineral.

They are also high in fibre, and thus have a detoxifying value. These are the same auspicious grains that re described in the Verdict texts and are used for Vedict ceremonies.

One-half cup of Dalia (measured dry), for instance, contains 14g. protein, 8mg. iron, and also magnesium and zine.The same amount of Rava contains 13g. of protein, 9.mg of iron and 3 mg. of zinc. Rye is also high in protein with one-half cup yielding 15 grams of protein and 4mg. of zine. Millet is a good source of B vitamins

All of these grains contain copper, which is an essential trace mineral that improves energy and immunity, and their zinc content also boosts ojas, the finest product of digestion that creates lightness, inner energy, immunity and bliss.

To prepare power grains, place two cups of water in a saucepan and bring it to a boi'. Add a teaspoon a Ghee and 1/2 cup of grain. Boil for then lower to a simmer. Cook is important because otherwise grains create internal dryness.

ENERGIZING VEGETABLES AND FRUITS

Other high-energy foods include fresh vegetables, which should constitute forty percent of the meal. Green, leafy vegetables are especially high in mineral and fibre, so should be eaten often.

Fruits are also a source of energy. You can start the day with a stewed apple, and if you feel hungry in between meals, try snacking on a juicy pear. If you are feeling heavy and bloated after lunch, eat a fresh papaya, it contain enzymes aiding digestion.

If you have a strong digestion and more Pitta in your constitution, mangoes are a rich ojas-producing food, and half a mango contains 2 mg beta-carotene and is a rich source of Vitamin

C. According to Ayurveda, Raisins are among the best of fruits because they enhance sattva (purity), pacify the mind and heart and increase the coordination between them.

They are also a rich source of iron and Vitamin B6, and provide magnesium, calcium, zinc and potassium. Raisins aid digestion and elimination when they are soaked in water overnight. One handful per person is a good amount. A Date-Milk Energy Shake is a nourishing way to end the day. Because it promotes sleep and calm both Pitta and Vata sleep imbalances.

Date-Milk Energy Shake

4-5 whole dates (Medjool dates are ideal)

1 cup whole organic milk (unhomogenized if possible)

1 pinch cinnamon powder

The important thing is to eat foods every day that boost your energy, rather than relying on artificial boosters when you feel your energy sag.

Boil the milk until creates foam. Turn off the heat and cool until the temperature is comfortable for drinking. Combine the milk with the other ingredients and blend until the dates are ground up. Drink it warm in winter and at room temperature is summer.

Beware of Energy-Draining Foods

Just as there is food to boost energy, other foods drain it. Any fast food, canned, frozen, packaged, leftover or old foods or foods laced with preservatives, chemicals, and additives-are difficult to digest and contain little nutritional content. If you do eat some of these foods, and you feel heavy after eating, Ayurveda advices drinking half a glass of water with 1/4 of a fresh lime squeezed into it.

But if you are feeling dull, sluggish, and drained of energy every day, it probably means that your diet contains too many energy-draining foods, which have clogged your micro channels with toxins. This mean that you need to overhaul your diet to include foods that create ojas and energy.

The important thing is to eat food every day that boost your energy, rather then relying on a artificial boosters when you feel your energy sag, because there will always be a negative side-effect.

FOOD AND CONCIOUSNESS

Attention Deficit Hyperactivity Disorder

What is ADHD?

ADHD is a diagnosis characterized by behaviours showing inappropriate levels of inattention, impulsivity and hyperactivity. These behaviours appear in school, at home and in social situations, and typically become worse in situations where sustained attention is required (eg. doing seat-work, studying, listening to people talk). The behaviours associated with ADHD usually become apparent before the age of four, but are frequently not recognized until the child commences school. It is reported that up to 6 percent of children may show symptoms of ADHD, with males outnumbering females approximately three-to-one.

Cautions:

- many of the behaviours/characteristics of ADHD are present in all children to some degree and at particular ages;
- the behaviours are not abnormal in themselves-only when excessive for the child's age.
- what characterizes ADHD is the intensity,persistence, and patterning of the behaviours which significantly impairs social, academic, and work activities.

Behaviours Associated with ADHD

1. Inattention:
   - often fails to give close attention to details; makes careless mistakes in schoolwork and other activities
   - often has difficulty sustaining attention in tasks or play activities
   - often does not seem to listen when spoken to directly

- often does not follow through on instructions and fails to finish schoolwork, chores, or other duties
- often has difficulty organizing tasks and activities
- often avoids, dislikes, or is reluctant to engage in tasks that require sustained mental effort (such as schoolwork or homework)
- often loses things necessary for tasks or activities (e.g. toys, school assignments, pencil, books, or tools)
- often easily distracted by extraneous stimuli
- often forgetful in daily activities

2. Impulsivity:
   - often blurts out answers before questions have been completed
   - often has difficulty awaiting turn
   - often interrupts or intrudes on others
3. Hyperactivity:
   - often fidgets with hands or feet or squirms in seat
   - often leaves seat in the classroom or wherever remaining seated is expected
   - often runs about or climbs excessively in situations where it is inappropriate
   - often has difficulty playing or engaging in leisure activities quietly
   - often "on the go" or often acts as if "driven by a motor"
   - often talks excessively

Who can Diagnose ADHD?

Psychologists, psychological associates (with access to diagnosis) and physicians properly diagnose ADHD using criteria from the three characteristics of inattention, impulsivity and hyperactivity. A child must show a significant number of the characteristic behaviours more frequently than expected for a child that age, in at least two different environments, and for at least six consecutive months. Onset of these behaviours should be before the age of seven. Since a child may show difficulties with sustaining attention for a variety of reasons, psychologists diagnosing ADHD will use a multidisciplinary approach involving parents, classroom teachers and other individuals who know the child.

How is ADHD Best Treated?

There is no single effective method for treating ADHD. Research in modifying children's environments; counselling; behaviour modification; medication and individualized academic programing all have shown promise. No one technique, nor single combination of techniques is effective with all children. In particular, treating ADHD with stimulant medication remains inconclusive. There is little evidence that academic achievement improves with the use of medications such as Ritalin, Dexedrine and Cylert alone. Medication can and does increase the attention span of children. However, it is unclear that thinking, memory and academic skills are improved by simply helping children to attend and engage in learning activities for longer periods of time. Similarly, other treatment approaches to ADHD which emphasize specific diets, megavitamins, restrictions on food dyes and listening stimulation, have not been shown to have a significant effect on ADHD in well controlled research studies.

In the midst of all the uncertainty as to how to deal effectively with ADHD children, it appears that an eclectic or broad-based approach is best. A combination of behaviour modification, medication, and classroom and home management techniques has shown the most promising results. Only an individualized approach implemented by parents and teachers, under the direction and supervision of psychologists and physicians, will meet the special needs of children with ADHD.

Helpful Tips:

- only experienced psychologists and physicians can accurately diagnose ADHD.
- parents should be cautious not to diagnose their children after reading the characteristic behaviours of ADHD (be aware and concerned, but don't diagnose).

- seek advice from an experienced psychologist or physician for an accurate diagnosis and appropriate treatment plan.
- all methods of treatment should be closely monitored; be wary of the "take these pills and see me in six months" approach.

Facts and Figures about ADHD

According to the National Institutes of Mental Health (NIMH), "ADHD (Attention Deficit Hyperactivity Disorder) refers to a family of related chronic neurobiological disorders that interfere with the individual's capacity to regulate activity level (hyperactivity), inhibit behaviour (impulsivity), and attend to tasks (inattention) in developmentally appropriate ways."

ADHD, under different labels, has been recognized as a disorder from at least the 1940's. Far from being an "invented disease", it is well substantiated that ADHD is a disorder which involves a dysfunction in the prefrontal cortex of the brain. This is the area of the brain which is involved in higher-order planning skills, and which permits us to inhibit responding. Using sophisticated neural imaging techniques, it has been discovered that there are structural differences in the brains of individuals with ADHD. The parts of the brain responsible for controlling attention have been found to use less glucose (a major fuel source) in ADHD individuals compared to nonaffected individuals. In addition, evidence strongly suggests that ADHD individuals also demonstrate neurochemical differences, in particular less of a chemical neurotransmitter called dopamine which is needed in some areas of the brain to focus attention and to control impulsive responding.

There is very strong evidence of a genetic component in this disorder, again refuting suggestions that the condition is not real. At least 1/3 of fathers who were diagnosed as having ADHD as children have children who are also diagnosed with ADHD. Siblings of ADHD children are 5 to 7 times more likely to demonstrate the disorder than are children from unaffected families. Even more compelling, the majority of identical twins share the trait.

Nongenetic causes account for about 20 to 30 percent of ADHD cases, and include fetal distress, premature birth, maternal alcohol use during pregnancy, exposure to high levels of lead during early childhood, traumatic brain injury (especially injury of the frontal lobes) and sleep problems. There is no evidence that ADHD is caused by food allergies, by watching too much TV, by excess sugar, by poor home life, or by poor schooling. Current data from the U.S. National Institutes of Mental Health (NIMH) estimate that about 5 to 6 percent of school age boys in the U.S. are diagnosed with this disorder. Boys with the disorder outnumber girls by a factor of 2 or 3. This translates into about 2 million children in the U.S. Similar percentages are reported for Canadian children.

There is no independent test for ADHD. This is a situation common to most psychiatric disorders, including schizophrenia and autism, due to their complexity and variability. Primary care and developmental pediatricians, family practitioners, child neurologists, psychologists and psychiatrists are the providers responsible for assessment, diagnosis and treatment of most children with ADHD. NIMH data indicate that family doctors diagnose more quickly and prescribe medication more frequently than psychiatrists and pediatricians. Misdiagnosis can occur, as other disorders (learning disabilities, chronic ear infections, language delays, adjustment problems, substance abuse, personality disorders, epilepsy) can produce similar symptoms.

In addition, a number of other disorders often accompany ADHD, including learning disabilities, fetal alcohol syndrome or effects, developmental delays, language disorders, anxiety disorders, mood disorders, conduct disorders and tic disorders, including Tourette's syndrome.

The more carefully the assessment is conducted, and the more information considered (e.g., medical, developmental, school, psychosocial and family histories, parent and teacher ratings, psychological testing) the lower the incidence is found to be for this disorder. Diagnosis should be based upon a thorough investigation of the child's life, in and outside of the home. Many children do not "outgrow" ADHD: about 1/3 of individuals diagnosed as children still meet the formal criteria as adults.

Hyperactive-impulsive symptoms generally decrease with age, but symptoms of inattention do not.

Without treatment, ADHD children are at significantly higher risk for a variety of negative life outcomes. They are more likely to be involved in accidents, especially motor vehicle accidents during adolescence, as well as tobacco use, early pregnancy and lower educational attainment. Almost half of ADHD children (mostly boys) exhibit symptoms of Oppositional Defiant Disorder, including stubbornness, bursts of temper, belligerent and defiant behaviour. In some of these children, there is a progression into the much more serious behavioural problems associated with a diagnosis of Conduct Disorder. About 25 percent of these children experience anxiety or depression. Some develop antisocial personality disorder.

There are three types of stimulant most commonly prescribed for this disorder, and Ritalin has proved to be the most effective. Its clinical effects peak about one hour after each dose, and dissipate in about 4 hours. Ritalin appears to act by increasing the amount of dopamine at the synapses, addressing the deficiency which may underlie this disorder. Nine out of ten children with ADHD improve with the use of one of the three stimulant medications.

Ritalin has been prescribed for children since the 1970's and consequently has been well studied. Side effects do include sleep problems, appetite loss and, in rare cases, facial tics. There may be a negative effect on growth rate, although ultimate height does not appear to be affected. Interestingly, the NIMH study found that the highest level of side-effects was reported for children receiving the placebo ("sugar pill") rather than the actual stimulant medication. Ritalin can be addictive when abused by adolescents and adults. However, long-term tolerance is rare as there is no need to increase the dosage to maintain clinical effects. Contrary to concerns about potential drug abuse, ADHD boys treated with stimulants are significantly less likely to abuse drugs and alcohol when older.

The NIMH and the US Department of Education recently published the results of the most comprehensive study of treatment effectiveness for ADHD ever undertaken. The Multimodal Treatment Study of Children with ADHD cost $11 million (US), and conclusively demonstrated that long-term combinations of psychosocial and stimulant treatments, as well as thorough and well-monitored medical management alone, were both significantly superior to intensive behavioural treatments alone and to treatments offered by most medical practitioners in the community. Stimulants were found to have 3-fold greater benefit on behaviour ratings than on ratings of attention, and the best effect on academics and social behaviour were observed when medication was maintained for a year or more. In contrast, behavioural parent training and behavioural interventions in the classroom setting had limited effects on hyperactivity, but large effects in reducing oppositionality, defiance and aggression.

American statistics indicate that production of Ritalin has increased about 500 percent since 1990. Health Canada estimates that prescriptions for Ritalin have increased 637 percent over the same period. Current estimates place the number of Canadian children taking Ritalin at about 2 to 3 percent, which represents at most 145,000 children in the country. No one denies that misuse and abuse of this drug occur.

However, authorities agree that the increase in prescriptions primarily represents the fact that children with the inattentive type of this disorder (predominantly girls) are now being identified in addition to children with either the hyperactive-impulsive type and combined type, that children are being diagnosed earlier, and that they are staying on the drug for longer periods of time (often throughout adolescence and into adulthood). There are a number of treatments for ADHD which have no proven effectiveness: allergy treatments, megavitamins, chiropractic adjustment, treatment for yeast infection, eye training, or the use of specially coloured glasses. Restricted diets have been found in some studies to only help about 5 percent of children with ADHD, primarily younger children or those with food allergies.

Both Efalex and pycnogenol have no scientifically documented effectiveness in helping children with ADHD, and their manufacturers have been recently prohibited from making any such claims by unanimous decisions of the US Federal Trade Commission. For parents, the message should be clear: it does not help to delay effective treatment while trying out unknown or unproven ones. Families with ADHD kids have greater parental frustration, marital discord and rates of divorce. It is very hard to parent a child who is full of uncontrolled energy, who constantly leaves messes, throws tantrums and doesn't listen or follow instructions at a level appropriate for his/her age. If children do not develop age-appropriate attention and inhibitory control, they continue to encounter the same problems and to respond in the same ineffective ways, even though they may know better and try very hard to change. Discipline (reasoning, scolding and especially hitting) doesn't work, as the children are not choosing to act in these negative ways.

Children exposed to such an approach quickly label themselves as "bad", and come to experience scolding as virtually the only kind of attention that they get. As a result, they develop a cycle of frustration, blame and anger with their parents. In addition, research has demonstrated that children imitate or model such aggressive behaviour from their parents, and are more likely to try it with other children. In fact, recent data from the Canadian National Longitudinal Study of Children and Youth have indicated that children who experience hostile or ineffective parenting are nine times more likely to demonstrate behaviour problems than are children exposed to positive parenting approaches. The best approach consists of positive, consistent parenting, modest goals and flexible structure. The poor prognosis for children with this disorder who fail to receive effective treatment underscores the importance of early identification and intervention for children with ADHD. The complex of hyperactivity, impulsivity and attentional problems is frequently observable in children as early as the preschool years. This combination of difficulties is strongly predictive of later behaviour problems.

However, inattention and impulsivity can follow from many kinds of problems at this age (e.g., language delays, recurrent ear infections), as well as from other sources such as natural rambunctiousness, or stressors such as divorce, neglect or inadequate child care. Consequently, diagnosis of ADHD in preschoolers should only be made cautiously by experts well trained in child neurobiological disorders. Medication for this very young age group is generally not recommended, as results of such treatment have not been thoroughly evaluated for effectiveness or safety. The NIMH study recommends that treatment should focus on structured play school, along with parent training and support. Aggressive and anti-social behaviour has been found to consolidate between the ages of 6 and 10, suggesting that this age group is a particularly important one to screen for children with this disorder. Information should be collected from both parents and teachers, and ideally a multi-disciplinary approach should be implemented, involving both health-related and school-based assessments. Unfortunately, because of different legislative and funding structures, there is currently a "disconnect" between assessments conducted in these different settings. In spite of its demonstrated effectiveness, medication alone is not necessarily the best treatment for all children. Treatment decisions should be based upon each child's individual needs, personal and medical history, research findings, and other relevant factors. Treatment combining medical management with behavioural approaches is more effective in addressing issues such as anxiety, academic performance, oppositionality, parent-child relationships and social skills. Children receiving combined treatments tend to require about 20 percent less medication.

The most common method of treatment for AD(H)D is the use of medication. There are three types of medication used in the treatment of AD(H)D: stimulants-(Ritalin, Dexedrine, Cylert,) tricyclic antidepressants-(Tofranil, Janimine, Norpramin, Pertofrane.) and antihypertensive drugs. Many books have been written on the subject reviewing the advantages and disadvantages as well as the contraindications for each of those medications. Talk to your doctor and stay informed. Remember everyone is special and unique what works wonders on some children, or adults, may be devastating to others.

BIBLIOGRAPHY

Acquaah, George: *Understanding Biotechnology: An Integrated and Cyber-Based Approach*, NJ, Pearson/Prentice Hall, 2004.

Ausubel, F.M.: *Current Protocols in Molecular Biology*, New York, John Wiley and Sons, 1989.

Bernauer, Thomas: *Genes, Trade, and Regulation: The Seeds of Conflict in Food Biotechnology*, Princeton, Princeton University. 2003.

Brandwein, P.F.: *Sourcebook for the Biological Sciences*, San Diego, Harcourt Brace JOvanovich, 1986.

Chrispeels, Maarten : *Plants, Genes and Crop Biotechnology*, Sudbury MA, Jones and Barlett Publishers, 2003.

Clark, J.M.: *Experimental Biochemistry*, New York, W.H. Freeman and Company, 1977.

Cox, Earl D. : *Beyond Humanity: Cyber Evolution and Future Minds*, Roackland, Charles River Media, 1996.

David Sadava: *Plants, Genes and Crop Biotechnology*, Sudbury MA, Jones and Barlett Publishers, 2003.

Dodds, John H.: *Plant Genetic Engineering*, New York, Cambridge University Press, 1985.

Fransman M, Junne G, Roobeek A: *The Biotechnology Revolution?*, Oxford, Blackwell, 1995.

Friedberg, E.C.: *DNA Repair*, New York, WH Freeman and Company, 1985.

Fumento, Michael: *Bioevolution: How Biotechnology is Changing Our World*, San Francisco, Encounter Books, 2003.

Gaskell G: *Biotechnology-the Making of a Global Controversy*, Cambridge, Cambridge University Press, 2002.

George S. Paul: *Beyond Humanity: Cyber Evolution and Future Minds*, Roackland, Charles River Media, 1996.

Goodman, D.C.: *From Farming to Biotechnology: A Theory of Agro-industrial Development*, Oxford, Blackwell, 1987.

Goodsell, David S.: *Bionanotechnology: Lessons From Nature*, Hoboken, Wiley-Liss, 2004.

King, R.B., Mihony, G., and Sheldon, J.K.: *Practical Environmental Bioremediation*, Lewis Publishers, Boca Raton, 1992.

Kurzweil, Ray: *The Age of Spiritual Machines*, New York, Penguin Books, 1999.

Miller, Henry I. : *The Frankenfood Myth: How Protest and Politics Threaten the Biotech Revolution*, New York, Praeger, 2004.

Moravec, Hans: *Mind Children: The Future of Robot and Human Intelligence*, Cambridge, Harvard University Press, 1988.

Morholt, E. : *Sourcebook for the Biological Sciences*, San Diego, Harcourt Brace JOvanovich, 1986.

Murray, David: *Seeds of Concern: The Genetic Manipulation of Plants*, Sydney, University of New South Wales, 2003.

Old, R.W. : *Principles of Gene Manipulation*, London, Blackwell Scientific Publications, 1989.

Parens, Erik: *Enhancing Human Traits: Ethical and Social Implications*, Washington, D.C.: Georgetown University Press, 1998.

Postman, Neil: *Technopoly: The Surrender of Culture to Technology*, New York, Vintage Books, 1992.

Primrose, S.B.: *Principles of Gene Manipulation*, London, Blackwell Scientific Publications, 1989.

Retzer, W.J.: *Biotechnology Workbook*, Englewood Cliffs, Prentice Hall, 1991.

Rifkin, Jeremy: *The Biotech Century*, New York, Penguin Putnam, 1998

Shetty, Kalidas: *Food Biotechnology*, New York, Dekker/CRC Press, 2005.

Smith, John E.: *Biotechnology*, Cambridge, Cambridge University Press, 2004.

Towle, Albert: *Modern Biology*, Austin, Holt, Rinehart and Winston, 1988.

Walden, Richard: *Genetic Transformation in Plants*, England, Open University Press, 1988.

INDEX

♦♦♦